T0174788

OPTICAL IMAGING DEVICES

New Technologies and Applications

Devices, Circuits, and Systems

Series Editor
Krzysztof Iniewski
CMOS Emerging Technologies Research Inc.,
Vancouver, British Columbia, Canada

PUBLISHED TITLES:

Atomic Nanoscale Technology in the Nuclear Industry
Taeho Woo

Biological and Medical Sensor Technologies
Krzysztof Iniewski

Building Sensor Networks: From Design to Applications
Ioanis Nikolaidis and Krzysztof Iniewski

Circuits at the Nanoscale: Communications, Imaging, and Sensing
Krzysztof Iniewski

CMOS: Front-End Electronics for Radiation Sensors
Angelo Rivetti

Design of 3D Integrated Circuits and Systems
Rohit Sharma

Electrical Solitons: Theory, Design, and Applications
David Ricketts and Donhee Ham

Electronics for Radiation Detection
Krzysztof Iniewski

**Embedded and Networking Systems:
Design, Software, and Implementation**
Gul N. Khan and Krzysztof Iniewski

Energy Harvesting with Functional Materials and Microsystems
Madhu Bhaskaran, Sharath Sriram, and Krzysztof Iniewski

**Graphene, Carbon Nanotubes, and Nanostuctures:
Techniques and Applications**
James E. Morris and Krzysztof Iniewski

High-Speed Devices and Circuits with THz Applications
Jung Han Choi

High-Speed Photonics Interconnects
Lukas Chrostowski and Krzysztof Iniewski

**High Frequency Communication and Sensing:
Traveling-Wave Techniques**
Ahmet Tekin and Ahmed Emira

FORTHCOMING TITLES:

Cell and Material Interface: Advances in Tissue Engineering, Biosensor, Implant, and Imaging Technologies
Nihal Engin Vrana

Circuits and Systems for Security and Privacy
Farhana Sheikh and Leonel Sousa

CMOS Time-Mode Circuits and Systems: Fundamentals and Applications
Fei Yuan

Electrostatic Discharge Protection of Semiconductor Devices and Integrated Circuits
Juin J. Liou

Gallium Nitride (GaN): Physics, Devices, and Technology
Farid Medjdoub and Krzysztof Iniewski

Ionizing Radiation Effects in Electronics: From Memories to Imagers
Marta Bagatin and Simone Gerardin

Mixed-Signal Circuits
Thomas Noulis and Mani Soma

Magnetic Sensors: Technologies and Applications
Simone Gambini and Kirill Poletkin

MRI: Physics, Image Reconstruction, and Analysis
Angshul Majumdar and Rabab Ward

Multisensor Data Fusion: From Algorithm and Architecture Design to Applications
Hassen Fourati

Nanoelectronics: Devices, Circuits, and Systems
Nikos Konofaos

Nanomaterials: A Guide to Fabrication and Applications
Gordon Harling, Krzysztof Iniewski, and Sivashankar Krishnamoorthy

Physical Design for 3D Integrated Circuits
Aida Todri-Sanial and Chuan Seng Tan

Power Management Integrated Circuits and Technologies
Mona M. Hella and Patrick Mercier

Radiation Detectors for Medical Imaging
Jan S. Iwanczyk

Radio Frequency Integrated Circuit Design
Sebastian Magierowski

Reconfigurable Logic: Architecture, Tools, and Applications
Pierre-Emmanuel Gaillardon

Silicon on Insulator System Design
Bastien Giraud

OPTICAL IMAGING DEVICES

New Technologies and Applications

EDITED BY

AJIT KHOSLA

UNIVERSITY OF CALGARY, ALBERTA, CA
LAB177 INC., ONTARIO, CA

DONGSOO KIM

SAMSUNG ELECTRONICS, SUWON, KOREA

KRZYSZTOF INIEWSKI MANAGING EDITOR

CMOS EMERGING TECHNOLOGIES RESEARCH INC.
VANCOUVER, BRITISH COLUMBIA, CANADA

CRC Press
Taylor & Francis Group
Boca Raton London New York

CRC Press is an imprint of the
Taylor & Francis Group, an **informa** business

CRC Press
Taylor & Francis Group
6000 Broken Sound Parkway NW, Suite 300
Boca Raton, FL 33487-2742

First issued in paperback 2020

© 2016 by Taylor & Francis Group, LLC
CRC Press is an imprint of Taylor & Francis Group, an Informa business

No claim to original U.S. Government works

ISBN-13: 978-1-4987-1099-2 (hbk)
ISBN-13: 978-0-367-77899-6 (pbk)

**Visit the Taylor & Francis Web site at
http://www.taylorandfrancis.com**

**and the CRC Press Web site at
http://www.crcpress.com**

Contents

Preface

Optical Imaging Devices: New Technologies and Applications provides an essential understanding of the design, operation, and practical applications of optical imaging and sensing systems, making it a handy reference for students and practitioners alike.

This book delivers a comprehensive introduction to optical imaging and sensing, from devices to system-level applications. Drawing upon the extensive academic and industrial experience of its prestigious editors and renowned chapter authors, this authoritative text explains the physical principles of optical imaging and sensing, providing an essential understanding of the design, operation, and practical applications of optical imaging and sensing systems.

It covers multidisciplinary topics such as fluorescence imaging system for freely moving animals, eye-in-hand robotics, smart eye tracking sensors, silicon-based imaging characteristics, nanophotonic phased arrays, thin-film sensors, label-free DNA sensors, and in vivo flow cytometry.

We would like to thank all the authors for their contributions to this book. Finally, we would like to thank the staff of CRC, especially Nora Konopka, for their kind help in publishing this book.

Editors

Dr. Ajit Khosla is a professor at the Soft & Wet Matter Engineering Lab (SWEL), Graduate School of Science and Engineering, Yamagata University, Japan, and a visiting professor at the MEMS Research Lab, Department of Mechanical Engineering, College of Engineering, San Diego State University, California, USA. His research work in the area of nano-microsystems has resulted in over ninety-five scientific and academic contributions. These contributions include: four keynote lectures, four patents, twenty journal papers, eight invited papers, two technical magazine articles, forty-five conference papers, twelve invited talks and one book chapter. His work has been referenced and downloaded over 1000 times since 2008. Dr. Khosla received a PhD from Simon Fraser University, Burnaby, British Columbia, Canada, where he was awarded the 2012 Dean of Graduate Studies Convocation Medal for his PhD research work on the development of novel micropatternable multifunctional nanocomposite materials for flexible nano- and microsystems. He then served as a postdoctoral fellow at Concordia University, Montreal, Québec, Canada, and at the University of Calgary, Alberta, Canada. In 2014, he founded and established a successful nano-microsystem company called Lab177 Inc. in Chatham, Ontario, Canada, which focuses on nano-microsystems in healthcare and flexible electronics. An active member in the scientific community, he is an Executive and Program Committee member for two major conferences, including Treasurer for the Electrochemical Society (ECS): Sensor Division, and the Society of Photographic Instrumentation Engineers (SPIE): Smart Structures and Materials-NDE. He has organized various Institute of Electrical and Electronics (IEEE) technical meetings and has been lead organizer for two major conference tracks at ECS, and also a guest editor for a focus issue on: (1) Microfluidics, MEMS/NEMS Sensors and Devices; (2) Nano-Microsystems in Healthcare and Environmental monitoring. He has served on the editorial board of the *Journal of Microelectronics Engineering* (Elsevier) since May 12th, 2013. Furthermore, he is also Editor for the yearly Special Issue with *Journal of Microsystem Technologies* (Springer) on Smart Systems.

Dongsoo Kim received his MS and PhD degrees in electrical and electronics engineering from Yonsei University, Seoul, Korea. Currently, Dr. Kim works as a principal engineer in the Wireless Communication Division at Samsung Electronics, Suwon, Korea. Previously, he was a staff analog design engineer at Aptina Imaging, San Jose, California, USA, and a postdoctoral associate in the Department of Electrical Engineering at Yale University, New Haven, Connecticut, USA. His research interests include CMOS image sensors, smart sensors, low-noise circuit design, and biomedical instrumentation.

Contributors

G. Alenyà
Institut de Robòtica i Informàtica
   Industrial
CSIC-UPC
Barcelona, Spain

Ahmed Arafa
University of British Columbia
Vancouver, British Columbia, Canada

S. Foix
Institut de Robòtica i Informàtica
   Industrial
CSIC-UPC
Barcelona, Spain

Yusaku Fujii
Department of Electronic Engineering
Gunma University
Kiryu, Gunma, Japan

Jinseok Heo
Department of Chemistry
Buffalo State College
Buffalo, New York
heoj@buffalostate.ed

Jonathan F. Holzman
University of British Columbia
Vancouver, British Columbia, Canada

Reynald Hoskinson
Department of Electrical and Computer
   Engineering
Department of Mechanical Engineering
University of British Columbia
Vancouver, British Columbia

Blago A. Hristovski
University of British Columbia
Vancouver, British Columbia, Canada

Xian Jin
University of British Columbia
Vancouver, British Columbia, Canada

Chang-Soo Kim
Departments of Electrical and
   Computer Engineering and
   Biological Sciences
Missouri University of Science and
   Technology
Rolla, Missouri

Dongsoo Kim
Department of Electrical and Electronic
   Engineering
Yonsei University
Seoul, South Korea
and
Department of Electrical Engineering
Yale University
New Haven, Connecticut

Mutsumi Kimura
Ryukoku University
Kyoto, Japan

Richard Klukas
University of British Columbia
Vancouver, British Columbia, Canada

Koichi Maru
Department of Electronics and
   Information Engineering
Kagawa University
Takamatsu, Kagawa, Japan

Joon Hyuk Park
Senior Produce Engineer
Arccos Golf
Stamford, Connecticut

Boris Stoeber
Department of Electrical and Computer
    Engineering
Department of Mechanical Engineering
University of British Columbia
Vancouver, British Columbia

Carme Torras
Institut de Robòtica i Informàtica
    Industrial
CSIC-UPC
Barcelona, Spain

Tao Wang
City College
City University of New York
New York, New York

Zhigang Zhu
City College
City University of New York
New York, New York

1 Thin-Film Sensors Integrated in Information Displays

Mutsumi Kimura

CONTENTS

Thin-film transistors (TFTs) have been widely utilized for flat-panel displays (FPDs). The essential feature is that functional devices can be fabricated on large flexible substrates at low temperature at low cost. The outstanding advantage can be employed in sensor applications and maximized when they are integrated in FPDs. Here, first, we propose a p/i/n thin-film phototransistor as an excellent thin-film photodevice. Next, we compare two types of temperature sensor, 1-transistor 1-capacitor (1T1C) type and ring oscillator type, whose sensitivities are roughly 1°C. Finally, we also propose a magnetic field sensor using a micro-Hall device, which realizes real-time area sensing of a magnetic field.

1

1.1 INTRODUCTION

Thin-film transistors (TFTs) [1, 2] have been widely utilized for flat-panel displays (FPDs) [3], such as liquid crystal displays (LCDs) [4], organic light-emitting diode (OLED) displays [5], and electronic papers (EPs) [6]. They have been recently applied to driver circuits, system-on-panel units [7], and general electronics [8], such as information processors [9], integrated in the FPDs. However, the essential feature of the thin-film technology is that functional devices, such as active-matrix circuits, amplifying circuits, and general circuits, can be fabricated on large and flexible substrates at low temperature at low cost. This outstanding advantage can be employed in not only the FPDs, but also sensor applications because the active-matrix circuits are required for some kinds of area sensors, amplifying circuits are necessary to amplify weak signals from sensor devices, large areas are sometimes needed to improve the sensitivities, and flexible substrates are convenient when they are put anywhere on demand. Moreover, this advantage can be maximized when they are integrated in the FPDs because the thin-film sensors and TFTs can be simultaneously manufactured using the compatible fabrication process. Therefore, we have studied potential possibilities of the thin-film sensors based on TFT technologies [10].

In this article, as examples of the thin-film sensors, we introduce a photosensor, temperature sensor, and magnetic field sensor. First, we propose a p/i/n thin-film phototransistor (TFPT) [11], compare it with other photodevices [12], characterize it from the viewpoint of the device behavior [13], and conclude that the p/i/n TFPT is an excellent thin-film photodevice. Examples for the application of the thin-film photodevices integrated in FPDs are a peripheral light sensor to control brightness of FPDs and an optical touchpanel. The p/i/n TFPTs are also promising for artificial retinas [14–21], which is not explained here. Next, we contrive two types of temperature sensor, 1-transistor 1-capacitor (1T1C) type [22] and ring oscillator type [23], and compare them. It is known that the TFTs have temperature dependence of the current-voltage characteristic. We employ the temperature dependence to realize the temperature sensor, and invent a sensing circuit and sensing scheme. It is confirmed that the temperature sensitivities of these temperature sensors are less than 1°C. The 1T1C type temperature sensor has a simple circuit configuration, but it is needed to equip analog voltage sensing circuits, which are fairly complicated circuits and difficult to be integrated on the FPDs using TFTs. On the other hand, the ring oscillator type temperature sensor is available as a digital device. Generally, digital devices are superior to analog devices from the viewpoint of the onward signal operation. Examples for the application of the temperature sensors integrated in FPDs are a temperature sensor to optimize driving conditions and compensate temperature dependences of display characteristics of liquid crystals or organic light-emitting diodes, a thermal touchpanel, and a daily medical health check system. Finally, we also propose a magnetic field sensor using a micro-Hall device based on the Hall effect in the magnetic field [24], which realizes real-time area sensing of the magnetic field [25]. Although there remains room for improvement of the sensitivity, no other devices can realize real-time area sensing of the magnetic field. An example for the application of the magnetic field sensors integrated in FPDs is a terrestrial magnetism detector to know the bearing. Figure 1.1 shows the application of the thin-film sensors integrated in the FPDs.

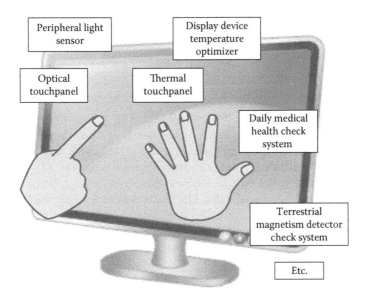

FIGURE 1.1 Application of the thin-film sensors integrated in the FPDs.

1.2 P/I/N THIN-FILM PHOTOTRANSISTOR

1.2.1 FABRICATION PROCESSES AND DEVICE STRUCTURES

Figure 1.2 shows the device structure of the p/i/n TFPT. The p/i/n TFPT is fabricated on a glass substrate using the same fabrication processes as low-temperature poly-Si (LTPS) TFTs [26–28]. First, an amorphous-Si film is deposited using low-pressure chemical vapor deposition (LPCVD) of Si_2H_6 and subsequently crystallized using a XeCl excimer laser to form a 50 nm thick poly-Si film. Next, a SiO_2 film is deposited using plasma-enhanced chemical vapor deposition (PECVD) of tetraethylorthosilicate (TEOS) to form a 75 nm thick control insulator film. A metal film is deposited and patterned to form a control electrode. Afterward, phosphorous ions are implanted through a photoresist mask at 55 keV with a dose of 2×10^{15} cm^{-2} to form an n-type anode region, and boron ions are implanted through a photoresist mask at 25 keV with a dose

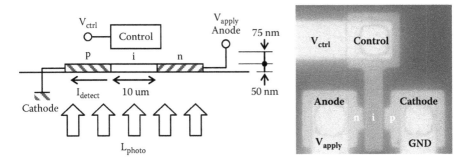

FIGURE 1.2 Device structure of the p/i/n thin-film phototransistor.

of 1.5×10^{15} cm^{-2} to form a p-type cathode region. Finally, a water vapor heat treatment is performed at 400°C for 1 h to thermally activate the dopant ions and simultaneously improve the poly-Si film, the control insulator film, and their interfaces.

1.2.2 Comparison of Thin-Film Photodevices

Thin-film photodevices generate the detected current (I_{detect}) caused by the generation mechanism of electron-hole pairs at semiconductor regions where the photoilluminance (L_{photo}) is irradiated, the depletion layer is formed, and the electric field is applied. It is of course preferable that the I_{detect} and corresponding photosensitivity, $\Delta I_{detect}/\Delta L_{photo}$, are high. It is also preferable that the I_{detect} is dependent only on the L_{photo}, but independent of the applied voltage (V_{apply}), because thin-film photodevices are mostly used to detect the L_{photo}, and the V_{apply} often unintentionally changes in the circuit configuration.

First, the characteristic of the p/i/n thin-film photodiode (TFPD) is shown in Figure 1.3a. It is found that the I_{detect} is proportional to the L_{photo} and relatively high

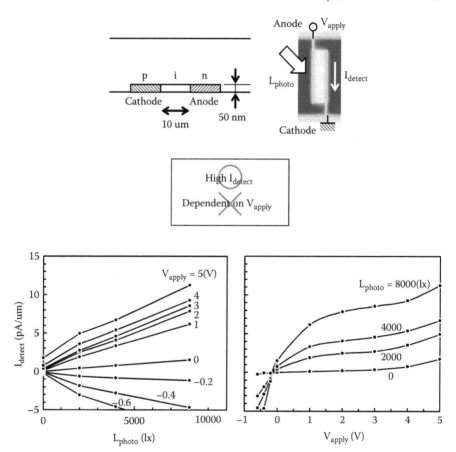

FIGURE 1.3 Comparison of the thin-film photo devices. (a) p/i/n thin-film photodiode.
(continued)

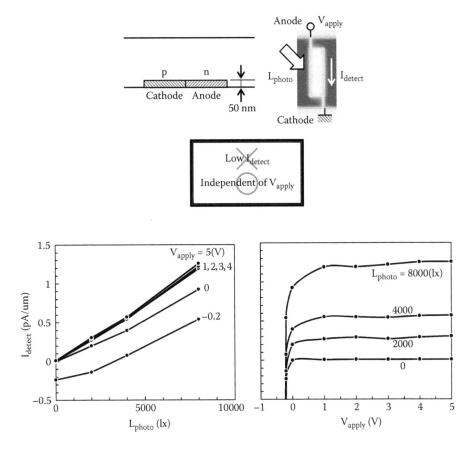

FIGURE 1.3 (CONTINUED) Comparison of the thin-film photo devices. (b) p/n thin-film photodiode.   *(continued)*

but dependent on the V_{apply}. This is because the depletion layer is formed in the whole intrinsic region, which means that the generation volume of electron-hole pairs is large, and the I_{detect} is relatively high. However, the electric field increases as the V_{apply} increases, which means that the generation rate increases, and the I_{detect} is dependent on the V_{apply}.

Next, the characteristic of the p/n TFPD is shown in Figure 1.3b. It is found that the I_{detect} is not high but independent of the V_{apply}. This is because the depletion layer is formed only near the p/n junction due to a built-in potential, which means that the generation volume of electron-hole pairs is small, and the I_{detect} is not high. However, the electric field is always high, which means the generation rate is constant, and the I_{detect} is independent of the V_{apply}.

Finally, the characteristic of the p/i/n TFPT is shown in Figure 1.3c, where the control voltage (V_{ctrl}) is optimized. It is found that the I_{detect} is simultaneously relatively high and independent of the V_{apply}. This is because the depletion layer is formed in the whole intrinsic region, which means that the generation volume of electron-hole pairs is large, and the I_{detect} is relatively high. Moreover, the electric field is

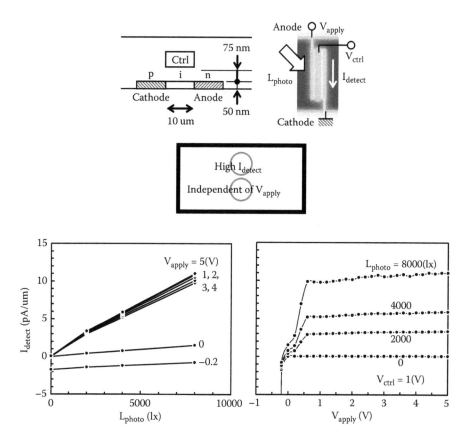

FIGURE 1.3 (CONTINUED)   Comparison of the thin-film photo devices. (c) p/i/n thin-film phototransistor.

always high because it is mainly determined by the V_{ctrl}, which means that the generation rate is constant, and the I_{detect} is independent of the V_{apply}. Therefore, the p/i/n TFPT is recommended.

1.2.3   DEVICE CHARACTERIZATION OF THE P/I/N THIN-FILM PHOTOTRANSISTOR

Figure 1.4a shows the optoelectronic characteristic of the actual device, namely, the measured dependence of I_{detect} on V_{ctrl} with a constant V_{apply} while varying L_{photo}. First, it is found that the dark current, I_{detect} when $L_{photo} = 0$, is sufficiently small except when V_{ctrl} is large. This is because the p/i junction and i/n junction act as insulators, preventing any significant electric current under the reverse bias. This characteristic is useful to improve the signal-to-noise (S/N) ratio of the p/i/n TFPT for photosensor applications. Next, I_{detect} increases as L_{photo} increases. This characteristic should be useful in quantitatively determining L_{photo}. Finally, I_{detect} is maximized when $V_{ctrl} \cong V_{apply}$. This relation is valid also for other V_{apply} [29]. This reason is discussed below by comparing an actual device with device simulation.

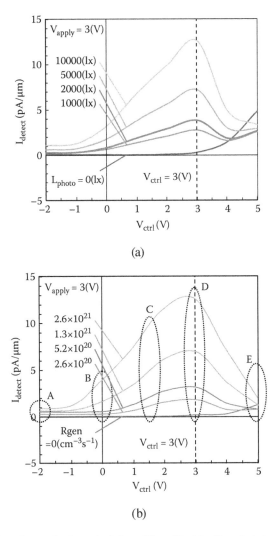

FIGURE 1.4   Optoelectronic characteristics of the p/i/n thin-film phototransistor. (a) Actual device. (b) Device simulation.

The same device structure as the actual device is constructed in the two-dimensional device simulator [30]. The uniform photoinduced carrier generation rate (R_{gen}) is set to 2.6×10^{17} cm^{-3}s^{-1}lx^{-1} as a fitting parameter by fitting the simulated I_{detect} to the experimental one. Schockley-Read-Hall recombination [31, 32], phonon-assisted tunneling with the Poole-Frenkel effect [33], and band-to-band tunneling [34] are included as generation-recombination models, as these models sufficiently simulate poly-Si thin-film devices [35]. The electron and hole mobilities are respectively set to 165 and 55 cm^{2}V^{-1}s^{-1}, which are determined from a transistor characteristic of a LTPS TFT fabricated at the same time. I_{detect} is calculated by sweeping V_{ctrl}, while maintaining V_{apply}, but changing R_{gen}.

Figure 1.4b shows the optoelectronic characteristic from the device simulation, namely, the calculated dependence of I_{detect} on V_{ctrl} with a constant V_{apply} while varying R_{gen}. The qualitative features of the experimental results are reproduced in the simulation results. In particular, I_{detect} is maximized when $V_{ctrl} \cong V_{apply}$. This reason is discussed below by considering the detailed behavior of the depletion layer and the difference between the electron and hole mobilities.

Figure 1.5a shows the electric potentials (φ) in the p/i/n TFPT, namely, the calculated distributions of φ in both the poly-Si film and the control insulator film in a cross

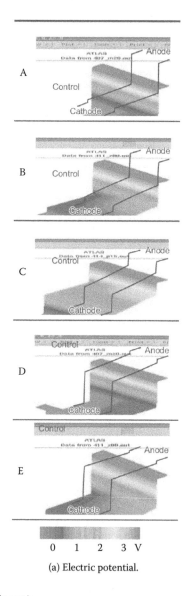

(a) Electric potential.

FIGURE 1.5   (See color insert.)   *(continued)*

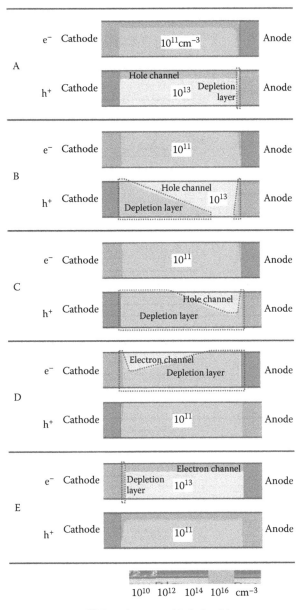

10^{10} 10^{12} 10^{14} 10^{16} cm^{-3}

(b) Free electron and hole densities.

FIGURE 1.5 (CONTINUED) (See color insert.) *(continued)*

section perpendicular to p/i/n TFPT as three-dimensional images. Figure 1.6 shows φ at the top poly-Si surface. φ is zero in the cathode electrode, which is grounded as shown in Figure 1.2. Thus, φ in the anode region and φ in the cathode region are not V_{apply} and zero owing to the built-in potentials in the n-type anode and p-type cathode

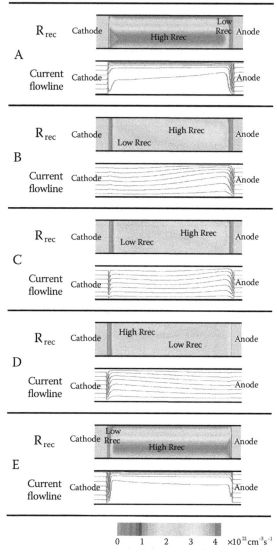

(c) Carrier recombination rate and current flowline

FIGURE 1.5 (CONTINUED)

regions, respectively. Figure 1.5b shows the free electron density (e⁻) and the free hole density (h⁺) in the poly-Si film, namely, the calculated distributions of e⁻ and h⁺ in a cross section perpendicular to the poly-Si film. The depletion layers are defined as regions where both free carrier densities are below 10^{12} cm⁻³. Figure 1.5c shows the carrier recombination rates (R_{rec}) and current flowline in the poly-Si film, namely, the calculated distributions of R_{rec} and current flowline in a cross section perpendicular to the poly-Si film. Although L_{photo} is assumed to be 10,000 lx in Figures 1.5 and

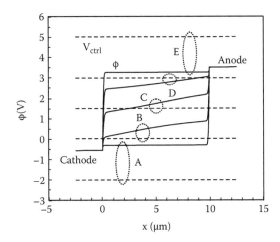

FIGURE 1.6 Electric potential at the top poly-Si surface.

1.6, the values of φ, e^-, h^+, and R_{rec} are qualitatively similar for all L_{photo}. The bias points A, B, C, D, and E in Figure 1.4b correspond to the graphs and figures A, B, C, D, and E in Figures 1.5 and 1.6, respectively.

I_{detect} depends on the volume of the depletion layer. R_{gen} is uniform in the poly-Si film, whereas R_{rec} is dependent on e^- and h^+. In the depletion layer, the electrons and holes are generated owing to light irradiation, and they do not recombine, but are transported and contribute to I_{detect} because both e^- and h^+ are low and the chance of the recombination is low. However, electrons and holes not in the depletion layer immediately recombine owing to the aforementioned generation-recombination models and do not contribute to I_{detect} because either e^- or h^+ is high and the chance of recombination is high. This is consistent with the presence of a depletion layer in Figure 1.5b that corresponds to the area where R_{rec} is low in Figure 1.5c. Consequently, the value of I_{detect} depends on the volume of the depletion layer.

A: When $V_{ctrl} < 0$, because $V_{ctrl} < \varphi$ throughout the intrinsic region, a hole channel is induced, and a pseudo-p/n junction appears near the anode region. Because a depletion layer is narrowly formed, I_{detect} is small. The current flowline indicates that the electric current generated equally within the depletion layer mainly flows through the hole channel.

B: When $V_{ctrl} \cong 0$, similar to the pinch-off phenomena in the saturation region of metal-oxide-semiconductor field-effect transistors (MOSFETs), although the hole channel is induced, because $V_{ctrl} \cong \varphi$ near the cathode region, h^+ decreases. Besides the depletion layer at the pseudo-p/n junction near the anode region, because another depletion layer is widely formed from the cathode region to the back surface, I_{detect} increases. The current flowline indicates that the electric current generated equally within the depletion layer relatively uniformly flows through the poly-Si film.

    C: When $0 < V_{ctrl} < V_{apply}$, because $V_{ctrl} > \varphi$ on the side of the cathode region, h$^+$ further decreases, but because V_{ctrl} remains below the threshold voltage, an electron channel is not observed. At the same time, because $V_{ctrl} < \varphi$ on the side of the anode region, the hole channel is induced. Because the depletion layer is widely formed from the cathode region to the underside of the hole channel, I_{detect} further increases. The current flowline still indicates that the electric current generated equally within the depletion layer relatively uniformly flows through the poly-Si film.

    D: When $V_{ctrl} \cong V_{apply}$, although an electron channel is induced, because $V_{ctrl} \cong \varphi$ near the anode region, e$^-$ is low. Because the depletion layer is widely formed from the anode region to the underside of the electron channel, I_{detect} further increases. The current flowline still indicates that the electric current generated equally within the depletion layer relatively uniformly flows through the poly-Si film. Incidentally, I_{detect} for B is smaller than I_{detect} for D. In the case of B, the generated holes cannot be rapidly transported because the hole mobility is less than the electron mobility; hence, the holes accumulate in the poly-Si film. On the other hand, in the case of D, the generated electrons can be rapidly transported because the electron mobility is high; hence, the electrons do not accumulate much in the poly-Si film. As a result, the depletion layer in B is less than that in D. Moreover, the value of I_{detect} depends on the volume of the depletion layer as well as e$^-$ and h$^+$. Consequently, I_{detect} for B is smaller than I_{detect} for D.

    E: When $V_{ctrl} > V_{apply}$, because $V_{ctrl} > \varphi$ throughout the intrinsic region, an electron channel is induced, and a pseudo-p/n junction appears near the cathode region. Because another depletion layer is narrowly formed, I_{detect} is small. The current flowline indicates that the electric current generated equally within the depletion layer mainly flows through the electron channel.

1.3 1-TRANSISTOR 1-CAPACITOR TYPE TEMPERATURE SENSOR

1.3.1 FABRICATION PROCESSES AND DEVICE STRUCTURES

The fabrication processes for the top-gate, coplanar, and n-type poly-Si TFTs are as follows [36–41]. First, an amorphous-Si film is deposited using LPCVD of Si_2H_6 and crystallized using a XeCl excimer pulse laser to form a poly-Si film, whose thickness (t_{Si}) is 50 nm. Next, a SiO_2 film is deposited using PECVD of TEOS to prepare a gate insulator film, whose thickness (t_{SiO2}) is 75 nm. Afterward, a gate metal film is deposited and patterned, and phosphorus ions are implanted and thermally activated to form source and drain regions. Subsequently, a SiO_2 film is deposited and patterned to prepare an interlayer insulator film, and a source and drain metal film is deposited and patterned. Finally, water vapor heat treatment is performed to improve the poly-Si film, the SiO_2 film, and their interfaces. The gate width (W) and length (L) are 50 and 4.5 μm, respectively. The field effect mobility (μ) and threshold voltage (V_{th}) are 93 cm$^2 \cdot$V$^{-1} \cdot$s^{-1} and 3.6 V, respectively.

1.3.2 TEMPERATURE DEPENDENCES OF TRANSISTOR CHARACTERISTICS

We measure the temperature dependence of the transistor characteristic. The poly-Si TFT is located on a chuck stage of a manual probe in a shield chamber, whose temperature is controlled by heating or cooling the chuck stage and measuring the temperature by a thermometer inserted into the chuck stage. Figure 1.7 shows the temperature dependence of the transistor characteristic of the poly-Si TFT. Although the transfer characteristic for the drain voltage $(V_{ds}) = 0.1$ V is only shown here, that for higher V_{ds} also has similar features. The temperature dependence of the off current is much larger than that of the on current. Moreover, although it is known that the off current of poly-Si TFTs is caused by multiple complicated mechanisms such as Schockley-Read-Hall generation (SRH), phonon-assisted tunneling with the Poole-Frenkel effect (PAT), and band-to-band tunneling (BBT) [42], the temperature dependence is monotonic, increasing from –30°C to 200°C. Furthermore, while the on current may affect device temperature owing to self-heating effect, the off current does not. Therefore, we determine that it is suitable to employ the off current for the temperature sensor.

1.3.3 CELL CIRCUIT AND DRIVING METHOD

We suggest a cell circuit consisting of 1-transistor and 1-capacitor and driving method composed of initializing, holding, and detecting periods. This sensing scheme can effectively utilize the temperature dependence of the off current. Figure 1.8 shows the cell circuit and driving method of this sensing scheme. Although the cell circuit is similar to the pixel circuit of LCDs, the cell capacitor (C_{cell}) has to have much larger capacitance by considering the actual working (see the next section). Such large capacitance can be located because the cell circuit does not have to be fine, unlike LCDs.

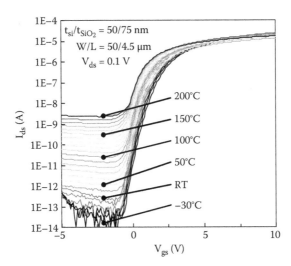

FIGURE 1.7 (See color insert.) Temperature dependence of the transistor characteristic of the poly-Si TFT.

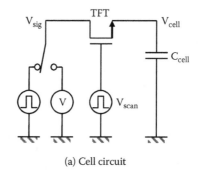

(a) Cell circuit

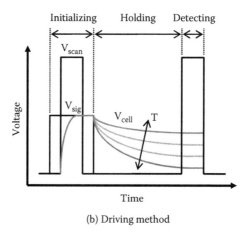

(b) Driving method

FIGURE 1.8 Cell circuit and driving method.

First, during an initializing period, scan voltage (V_{scan}) is applied in order to switch on the TFT, and signal voltage (V_{sig}) is applied. Then, cell voltage (V_{cell}) is charged and becomes equal to V_{sig}. Next, during a holding period, V_{scan} is grounded in order to switch off the TFT, and V_{sig} is also grounded. Then, V_{cell} is discharged and decreases toward V_{sig} due to the off current. Since the discharging speed depends on the temperature due to the temperature dependence of the off current, V_{cell} at the end of the holding period (V_{out}) also depends on the temperature. Finally, during a detecting period, V_{scan} is again applied, then V_{out} is detected through V_{sig}. As a result, it is possible to detect the temperature by measuring V_{out}.

1.3.4 EXPERIMENTAL RESULTS

We find that it is possible to detect the temperature using this sensing scheme, for example, from the room temperature to 70°C. Figure 1.9a shows the voltage waveform of this sensing scheme. Here, the TFT is the same as that written in the previous section, and C_{cell} is 100 pF. The initializing and holding periods are 200 μs and 20 ms, respectively. Pulse voltages of 10 and 5 V are applied as V_{scan} and V_{sig},

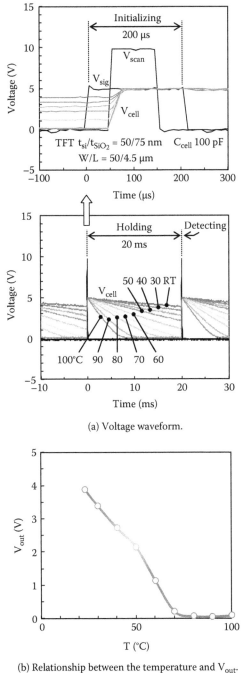

(a) Voltage waveform.

(b) Relationship between the temperature and V_{out}.

FIGURE 1.9 Experimental results.

respectively, and V_{cell} is measured using a high-impedance FET probe. The temperature is varied from room temperature to 100°C by considering the temperature range for regular working of the FET probe, although the temperature controller is the same as in the previous section. It is confirmed that the voltage waveform does not change even after the actual operation during several days, which suggests that the characteristic degradation of the poly-Si TFT does not occur even at the high temperature. However, further AC and DC stress tests are required to check long-time reliability, although the characteristic degradation of the poly-Si TFT hardly occurs at room temperature [43, 44]. Figure 1.9b shows the relationship between the temperature and V_{out}. It is possible to detect the temperature from room temperature to 70°C by measuring V_{out}.

Although this relationship might be uneven owing to the nonuniformity of the transistor characteristic of the poly-Si TFT, the temperature can accurately be detected by preparing lookup tables between the temperature and V_{out} for each cell circuit. Although the lookup tables must be calibrated, only several points, such as every 10°C, are necessary because the relationship between the temperature and V_{out} is nearly straight, as shown in Figure 1.9b. Even if multiple cell circuits are integrated, they can be simultaneously calibrated by putting all of them in temperature chambers. Since some automatic calibration circuits can also be integrated, such additional steps are not impractical. Although the need for the calibration should be discussed after the evaluation of the nonuniformity of the transistor characteristic, since the lookup table must be certainly prepared owing to the nonlinearity of the temperature dependence, we think that the additional cost for the calibration can be within a permitted extent.

Consequently, the detection accuracy will be less than 1°C because the relationship is nearly straight, as shown in Figure 1.9b. The sensitivity of V_{out} on the temperature is roughly 80 mV/°C, which is sufficiently large to detect by common readout circuits, including those using TFTs. Moreover, it is expected to extend the temperature range by controlling the holding period, which we will report in the near future.

1.4 RING OSCILLATOR TYPE TEMPERATURE SENSOR

1.4.1 FABRICATION PROCESSES AND DEVICE STRUCTURES

The top-gate, coplanar, solid-phase crystallized (SPC), n- and p-type, self-aligned, and offset TFTs are fabricated [45, 46]. Figure 1.10 shows the fabrication processes and device structures of the self-aligned and offset TFTs. An amorphous Si film is deposited on a quartz substrate using LPCVD of SiH_4 and crystallized using furnace annealing in N_2 ambient to form a poly-Si film. A SiO_2 film is grown using thermal oxidation in O_2 ambient, and another SiO_2 film is stacked using chemical vapor deposition (CVD) to form a gate insulator film. The poly-Si film is upgraded using postannealing at 1000°C for 1 h. A metal film is deposited and patterned to form gate electrodes, and a photoresist is patterned using photolithography as implantation masks. Phosphorus and boron ions are implanted, whose dose densities are 2×10^{15} and 1×10^{15} cm^{-2}, respectively, and activated using furnace annealing to form source-drain regions. The edge locations of the gate and source-drain regions

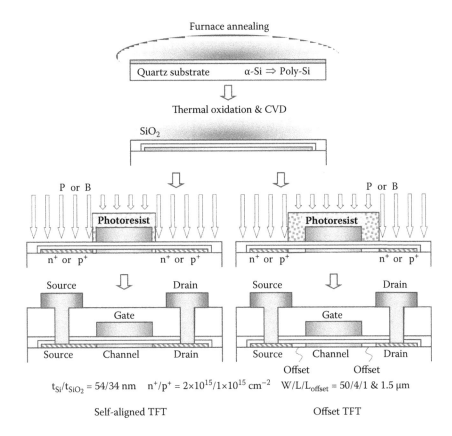

FIGURE 1.10 Fabrication processes and device structures of the self-aligned and offset TFTs.

in the self-aligned TFT are self-aligned, whereas those in the offset TFT are a certain length distant. The poly-Si film thickness (t_s) and gate insulator film thickness (t_i) are 54.0 and 33.7 nm, respectively. The gate width (W) and gate length (L) are 50 and 4 μm, respectively, and the offset length (L_{offset}) is 1.0 and 1.5 μm.

1.4.2 TEMPERATURE DEPENDENCES OF TRANSISTOR CHARACTERISTICS

The temperature dependences of the transistor characteristics are compared between the n- and p-type, self-aligned, and offset TFTs. Figure 1.11 shows the temperature dependences of the transistor characteristics. Here, the temperature (T) is varied from −20°C to 100°C. The vertical axes are different for each graph. The subthreshold swing (S), threshold voltage (V_{th}), and field effect mobility (μ) of the n-type self-aligned TFTs are 0.279 V·dec^{-1}, 1.03 V, and 67.3 cm^2·V^{-1}·s^{-1}, respectively, and those of the p-type self-aligned TFT are 0.215 V·dec^{-1}, −3.68 V, and 43.7 cm^2·V^{-1}·s^{-1}, respectively. It is confirmed that the temperature dependence of the n-type offset TFT is larger than that of the n-type self-aligned TFT. This is because the dominant mechanism of the on current in the self-aligned TFT is the channel conductance, which is hardly dependent on

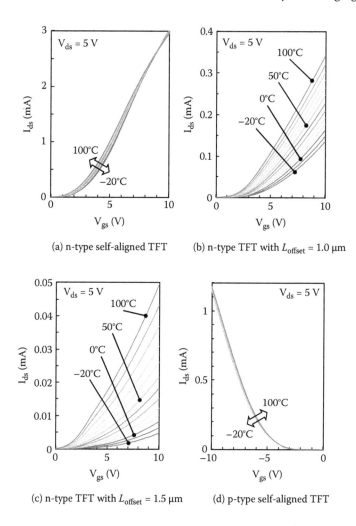

FIGURE 1.11   Temperature dependences of the transistor characteristics.

T, whereas that in the offset TFT is the conductance in the offset regions due to carrier generation-recombination processes [47], which seems to be strongly dependent on T. Moreover, the temperature dependence of the offset TFT with $L_{offset} = 1.5$ μm is larger than that of the offset TFT with $L_{offset} = 1.0$ μm. Therefore, it is preferable to use the offset TFTs with $L_{offset} = 1.5$ μm as a thermal sensor. As for the p-type TFTs, since the absolute value of the on current in the offset TFT is extremely small, which is not shown here, it is preferable to use the self-aligned TFT.

1.4.3   RING OSCILLATOR CIRCUIT

A static type ring oscillator is composed using the n-type offset TFTs with $L_{offset} = 1.5$ μm and p-type self-aligned TFTs, where the odd stages of complementary

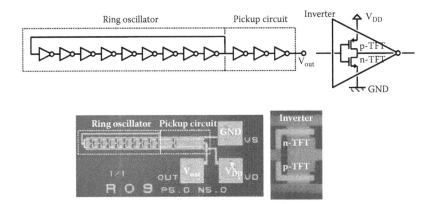

FIGURE 1.12   Circuit diagram and microscope photograph of the thermal sensor employing the ring oscillator.

metal-oxide-semiconductor (CMOS) inverters are circularly connected. Figure 1.12 shows the circuit diagram and microscope photograph of the ring oscillator. Here, W, L, and L_{offset} of the poly-Si TFTs in the CMOS inverters are 20, 5, and 1.5 µm, respectively. The supply voltage (V_{dd}) is applied, the output signal (V_{out}) is strengthened using the pickup circuit and observed using a high-impedance FET probe, and the oscillation frequency (f) is measured. It seems that f is mainly dependent on the on currents of the n- and p-type TFTs. Therefore, we expect that f is dependent on T.

1.4.4   EXPERIMENTAL RESULTS

The temperature dependences of f are measured. Figure 1.13 shows the temperature dependence of f for various V_{dd}. Here, T is varied from −20°C to 100°C. It is found

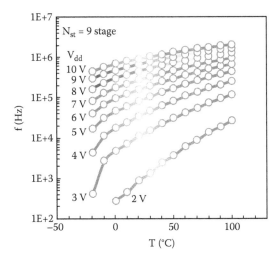

FIGURE 1.13   Temperature dependence of f for various V_{dd}.

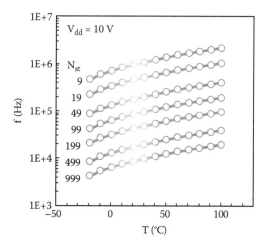

FIGURE 1.14 Temperature dependence of f for various N_{st}.

that f increases as T increases. Therefore, T can be detected by measuring f in principle. It is also found that f increases as V_{dd} increases. However, since V_{dd} is usually fixed to a certain value, the frequency range of f cannot be controlled by altering V_{dd}. Therefore, we consider altering the stage number of the CMOS inverters (N_{st}). Figure 1.14 shows the temperature dependence of f for various N_{st}. It is found that f decreases as N_{st} increases. Therefore, the frequency range of f can be controlled by altering N_{st} in order that f can be easily measured. In any case, it is clarified that T can be detected by measuring f.

1.5 MAGNETIC FIELD SENSOR

1.5.1 Matrix Array of Poly-Si Micro-Hall Devices

The matrix array of poly-Si micro-Hall devices is formed on a glass substrate using fabrication processes compatible with poly-Si TFTs [48–50] as follows. First, an amorphous-Si film is deposited using LPCVD of Si_2H_6, crystallized using a XeCl excimer pulse laser, and patterned to form poly-Si films. The laser wavelength is 308 nm, the irradiation intensity is 400 mJ/cm², the pulse duration is 28 nm, the shot number is 20 times, and the substrate temperature is room temperature. The poly-Si thickness is 50 nm, and the average grain size is 500 nm. Next, a SiO_2 film is deposited using PECVD of TEOS to form an implantation profile control film. The SiO_2 thickness is 75 nm. Phosphorous ions are then implanted with high- and low-dose densities of 2×10^{15} and 4×10^{13} cm⁻², to form leading lines and Hall channels, respectively. The acceleration voltage is 55 keV, and when the stopping power of the implantation profile control film is simulated [51] and the phosphorous ions are assumed to be uniformly profiled in the n-type poly-Si films, the doping densities are calculated to be 5.7×10^{19} and 1.1×10^{18} cm⁻³, respectively. Next, a SiO_2 film is deposited and patterned to form an encapsulation film, and a metal film is deposited and patterned to form contact electrodes. Finally, water vapor heat treatment is performed at 400°C for 1 h in order to activate the phosphorous ions, and to densify the SiO_2 films, while also improving the poly-Si and SiO_2 films, and their interfaces. The temperature is low enough that the average grain size is maintained. The electron densities for the high- and low-dose densities determined from the Hall measurements are 2.3×10^{19} and 4.6×10^{17} cm^{-3}, respectively, which means that the activation rate of the dopants is about 40%.

The matrix array of the poly-Si micro-Hall devices is shown in Figure 1.15. Here, 3×3 Hall devices are arrayed every 1×1 mm. Two current terminals and two Hall terminals are located perpendicular to each other. The width (W) and the length (L) of the Hall channels are 10 and 40 µm, respectively. The control voltage (V) is applied between the current terminals, and the control current (I) is measured between the current terminals. The magnetic field (B) is applied perpendicular to the Hall devices, and the Hall voltage (V_H) is measured between the Hall terminals. Although large contact pads are located around the Hall devices, the dimension of the principal part is less than 50×50 µm. This means that a high-resolution area sensor can be realized.

1.5.2 COMPENSATION TECHNIQUE OF CHARACTERISTIC VARIATION

The Hall voltages in the poly-Si micro-Hall devices have offset voltages (V_0) even when B is zero, whereas the changes in the Hall voltages are proportional to the magnetic field [52], i.e., $V_H = V_0 + (\partial V_H/\partial B) \cdot B$. The values of V_0 are different among the Hall devices because they are caused by I through the zigzag paths due to the random location of the polycrystalline grains, whereas the values of $\partial V_H/\partial B$ are similar among them. Although these phenomena are observed in actual measurements, the detailed mechanism has not been clarified and should be considered in theoretical discussions in the future. In spite of these issues, the poly-Si micro-Hall devices are superior to the nanocrystalline-Si and amorphous-Si micro-Hall devices because V_H is so large that it is proportional to the carrier velocity, i.e., carrier mobility.

The Hall voltages in the poly-Si micro-Hall devices are shown in Figure 1.16. Here, the Hall devices are located in uniform magnetic fields, and the dependences of V_H on B with a constant I are evaluated. Although only three sets of data from devices A, B, and C are shown, all data from 3×3 devices are evaluated in the same way. It is possible to detect unknown magnetic fields by measuring V_H and contrasting the dependences shown in Figure 1.16. However, as described above, it is found that the values of V_0 are different among the Hall devices, whereas the values of $\partial V_H/\partial B$ are similar among them. Therefore, a compensation technique of the characteristic variation is necessary. Each value of V_0 for each Hall device and a common value of $\partial V_H/\partial B$ for all Hall devices are calibrated in advance and stored, and V_H is converted into B using V_0 and $\partial V_H/\partial B$, i.e., $B = (V_H - V_0)/(\partial V_H/\partial B)$. The compensation technique is used in the following section.

The magnetic sensitivity, $\partial V_H/\partial B$, is 5 ~ 6 mV/T in this chapter, whereas it was 3 ~ 20 mV/T in a previous report [53]. This difference is owing to the differences of the device dimensions, carrier densities, driving conditions, etc., and indicates further possibility to improve the magnetic sensitivity.

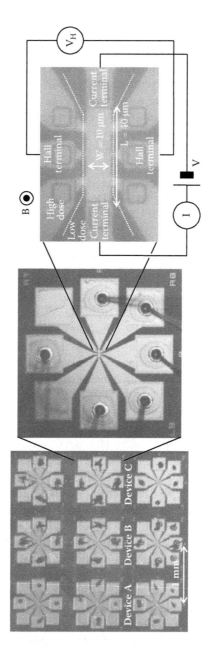

FIGURE 1.15 Matrix array of the poly-Si micro-Hall devices.

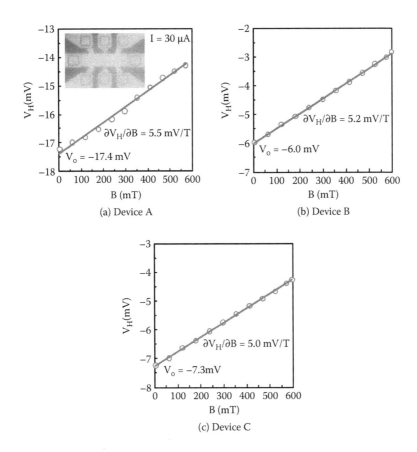

FIGURE 1.16   Hall voltages in the poly-Si micro-Hall devices.

1.5.3   AREA SENSING OF MAGNETIC FIELD

The area sensing of the magnetic field is shown in Figure 1.17. Here, a conventional Gauss meter is located in a nonuniform magnetic field generated intentionally, and the area sensing is conducted by moving the sensing probe every 1×1 mm. The matrix array of the poly-Si micro-Hall devices is also located in the same nonuniform magnetic field, and the area sensing is compared. The area sensing using the conventional Gauss meter and the matrix array are found to be roughly the same. Therefore, it is confirmed that the area sensing is correctly conducted using the matrix array. The magnetic resolution is less than 10 mT, as shown in Figure 1.17b. The response time is at most less than 0.1 s.

1.6   CONCLUSION

TFTs have been widely utilized for FPDs. The essential feature is that functional devices can be fabricated on large flexible substrates at low temperature by low cost. The outstanding advantage can be employed in sensor applications and maximized

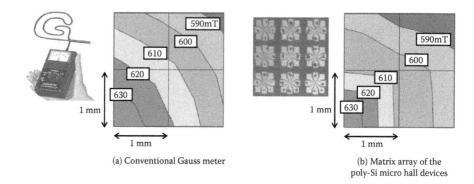

(a) Conventional Gauss meter

(b) Matrix array of the
poly-Si micro hall devices

FIGURE 1.17   (See color insert.) Area sensing of magnetic field.

when they are integrated in FPDs. Here, first, we proposed a p/i/n thin-film phototransistor as an excellent thin-film photodevice. Next, we compared two types of temperature sensor, the 1-transistor 1-capacitor (1T1C) type and the ring oscillator type, whose sensitivities are roughly 1°C. Finally, we also proposed a magnetic field sensor using a micro-Hall device, which realize real-time area sensing of the magnetic field.

REFERENCES

1. Y. Kuo, *Thin Film Transistors, Materials and Processes, 1: Amorphous Silicon Thin Film Transistors*, Kluwer Academic Publishers, Boston (2004).
2. Y. Kuo, *Thin Film Transistors, Materials and Processes, 2: Polycrystalline Silicon Thin Film Transistors*, Kluwer Academic Publishers, Boston (2004).
3. W. den Boer, *Active Matrix Liquid Crystal Displays: Fundamentals and Applications*, Newnes, Oxford (2005).
4. S. Morozumi, K. Oguchi, S. Yazawa, T. Kodaira, H. Ohshima, and T. Mano, *SID '83*, 156 (1983).
5. M. Kimura, I. Yudasaka, S. Kanbe, H. Kobayashi, H. Kiguchi, S. Seki, S. Miyashita, T. Shimoda, T. Ozawa, K. Kitawada, T. Nakazawa, W. Miyazawa, and H. Ohshima, *IEEE Trans. Electron Devices*, 46, 2282 (1999).
6. S. Inoue, H. Kawai, S. Kanbe, T. Saeki, and T. Shimoda, *IEEE Trans. Electron Devices*, 49, 1532 (2002).
7. Y. Aoki, and H. Kimura, *IDW '09*, 239 (2009).
8. Y. Kuo, *Jpn. J. Appl. Phys.*, 47, 1845 (2008).
9. N. Karaki, T. Nanmoto, H. Ebihara, S. Utsunomiya, S. Inoue, and T. Shimoda, *ISSCC '05*, 272 (2005).
10. M. Kimura, T. Hachida, Y. Nishizaki, T. Yamashita, T. Shima, T. Ogura, Y. Miura, H. Hashimoto, M. Hirako, T. Yamaoka, S. Tani, Y. Yamaguchi, Y. Sagawa, K. Setsu, Y. Imuro, and K. Bundo, *IMID '09*, 957 (2009).
11. T. Yamashita, T. Shima, Y. Nishizaki, M. Kimura, H. Hara, and S. Inoue, *Jpn. J. Appl. Phys.*, 47, 1924 (2008).
12. M. Kimura, T. Shima, T. Yamashita, Y. Nishizaki, and H. Hara, *IMID '07*, 7, 1745 (2007).
13. M. Kimura and Y. Miura, *IEEE Trans. Electron Devices*, 58, 3472 (2011).
14. T. Yagi, *Oyo Buturi*, 73, 1095 (2004) [in Japanese].
15. Y. Nakagawa, K.-W. Lee, T. Nakamura, Y. Yamada, K.-T. Park, H. Kurino, and M. Koyanagi, *ICONIP '00*, 636 (2000).

16. T. Noguchi, J. Y. Kwon, J. S. Jung, J. M. Kim, K. B. Park, H. Lim, D. Y. Kim, H. S. Cho, H. X. Yin, and W. Xianyu, *Jpn. J. Appl. Phys.*, 45, 4321 (2007).

17. J. Y. Kwon, H. Lim, K. B. Park, J. S. Jung, D. Y. Kim, H. S. Cho, S. P. Kim, Y. S. Park, J. M. Kim, and T. Noguchi, *Jpn. J. Appl. Phys.*, 45, 4362 (2007).

18. M. He, R. Ishihara, E. J. J. Neihof, Y. Andel, H. Schellevis, W. Metselaar, and K. Beenakker, *Jpn. J. Appl. Phys.*, 46, 1245 (2007).

19. H. Ueno, Y. Sugawara, H. Yano, T. Hatayama, Y. Uraoka, T. Fuyuki, J. S. Jung, K. B. Park, J. M. Kim, J. Y. Kwon, and T. Noguchi, *Jpn. J. Appl. Phys.*, 46, 1303 (2007).

20. T. Yamashita, T. Shima, Y. Nishizaki, M. Kimura, H. Hara, and S. Inoue, *Jpn. J. Appl. Phys.*, 47, 1924 (2008).

21. Y. Miura, T. Hachida, and M. Kimura, *IEEE Sensors J.*, 11, 1564 (2011).

22. A. Nakashima, Y. Sagawa, and M. Kimura, *IEEE Sensors J.*, 11, 995 (2011).

23. J. Taya, A. Nakashima, and M. Kimura, *Solid-State Electronics*, 79, 14 (2013).

24. Y. Yamaguchi, H. Hashimoto, T. Segawa, and M. Kimura, *Electrochem. Solid-State Lett.*, 14, J26 (2011).

25. Y. Yamaguchi, H. Hashimoto, M. Kimura, M. Hirako, T. Yamaoka, and S. Tani, *IEEE Electron Device Lett.*, 31, 1260 (2010).

26. S. Inoue, M. Matsuo, T. Hashizume, H. Ishiguro, T. Nakazawa, and H. Ohshima, *IEDM '91*, 555 (1991).

27. T. Sameshima, S. Usui, and M. Sekiya, *IEEE Electron Device Lett.*, 7, 276 (1986).

28. N. Sano, M. Sekiya, M. Hara, A. Kohno, and T. Sameshima, *IEEE Electron Device Lett.*, 16, 157 (1995).

29. M. Kimura, Y. Miura, T. Ogura, S. Ohno, T. Hachida, Y. Nishizaki, T. Yamashita, and T. Shima, *IEEE Electron Device Lett.*, 31, 984 (2010).

30. Silvaco International, Atlas, Device Simulator, http://www.silvaco.com/products/device_simulation/atlas.html

31. W. Shockley and W. T. Read Jr., *Phys. Rev.*, 87, 835 (1952).

32. R. N. Hall, *Phys. Rev.*, 87, 387 (1952).

33. O. K. B. Lui and P. Migliorato, *Solid-State Electronics*, 41, 575 (1997).

34. G. A. M. Hurkx, D. B. M. Klaassen, and M. P. G. Knuvers, *IEEE Trans. Electron Devices*, 39, 331 (1992).

35. M. Kimura, A. Nakashima, and Y. Sagawa, *Electrochem. Solid-State Lett.*, 13, H409 (2010).

36. H. Ohshima and S. Morozumi, *IEDM '89*, 157 (1989).

37. S. Inoue, M. Matsuo, T. Hashizume, H. Ishiguro, T. Nakazawa, and H. Ohshima, *IEDM '91*, 555 (1991).

38. T. Sameshima, S. Usui, and M. Sekiya, *IEEE Electron Device Lett.*, 7, 276 (1986).

39. H. Watakabe and T. Sameshima, *IEEE Trans. Electron Devices*, 49, 2217 (2002).

40. H. Watakabe, Y. Tsunoda, N. Andoh, and T. Sameshima, *J. Non-Crystalline Solids*, 299/302 B, 1321 (2002).

41. N. Sano, M. Sekiya, M. Hara, A. Kohno, and T. Sameshima, *IEEE Electron Device Lett.*, 16, 157 (1995).

42. M. Kimura and T. Tsujino, *IDW '07*, 1841 (2007).

43. N. Morosawa, T. Nakayama, T. Arai, Y. Inagaki, K. Tatsuki, and T. Urabe, *IDW '07*, 71 (2007).

44. T. Kasakawa, H. Tabata, R. Onodera, H. Kojima, M. Kimura, H. Hara, and S. Inoue, *IDW '09*, 1829 (2009).

45. M. Miyasaka, T. Komatsu, and H. Ohshima, *Jpn. J. Appl. Phys.*, 36, 2049 (1997).

46. H. Jiroku, M. Miyasaka, S. Inoue, Y. Tsunekawa, and T. Shimoda, *Jpn. J. Appl. Phys.*, 43, 3293 (2004).

47. A. Nakashima and M. Kimura, *IEEE Electron Device Lett.*, 32, 764 (2011).

48. S. Inoue, M. Matsuo, T. Hashizume, H. Ishiguro, T. Nakazawa, and H. Ohshima, *IEDM '91*, 555 (1991).

49. T. Sameshima, S. Usui, and M. Sekiya, *IEEE Electron Device Lett.*, 7, 276 (1986).

50. N. Sano, M. Sekiya, M. Hara, A. Kohno, and T. Sameshima, *IEEE Electron Device Lett.*, 16, 157 (1995).

51. J. F. Ziegler, J. P. Biersack, and M. D. Ziegler, SRIM—The Stopping and Range of Ions in Matter, http://www.srim.org/

52. M. Kimura, Y. Yamaguchi, H. Hashimoto, M. Hirako, T. Yamaoka, and S. Tani, *Electrochem. Solid-State Lett.*, 13, J96 (2010).

53. F. Le Bihan, E. Carvou, B. Fortin, R. Rogel, A. C. Salaün, and O. Bonnaud, *Sensors Actuators A*, 88, 133 (2001).

2 Analytical Use of Easily Accessible Optoelectronic Devices
Colorimetric Approaches Focused on Oxygen Quantification

Jinseok Heo and Chang-Soo Kim

CONTENTS

2.1 INTRODUCTION

Rapid technological progress in digital color imaging devices, such as charge-coupled devices (CCDs), complementary metal-oxide-semiconductor (CMOS) cameras, liquid crystal display (LCD), and digital light projection (DLP) devices, makes these optoelectronic products nearly ubiquitous in our daily lives. For example, relatively high-quality color images can be obtained anytime with mobile phone cameras and wireless webcams (some products as low as several tens of U.S. dollars) and then transmitted wirelessly. These color imaging devices were originally designed as sensory or perceptual devices that mimic the human eye responding to visible wavelengths. This chapter introduces the potential use of these color imaging devices as economic analytical instruments. After we briefly review prospective luminophores amenable for colorimetric chemical quantification using color imaging devices, we describe oxygen quantification as an exemplary application of this approach.

2.1.1 COLOR OPTOELECTRONIC DEVICES

The most widely used instruments for optical interrogation of biochemical agents are spectrometric systems that detect emission, absorption, or reflection signals from samples; however, they are generally expensive and bulky to use for cost-effective, portable sensors. Therefore, many lab-on-a-chip or batch-fabricated spectrometers have been actively developed recently for their common use as economic analytical instruments (Syms, 2009; Babin et al., 2009).

In parallel with this approach, the color image sensor is considered an alternative device to spectrometry because of its photometric detection capability (Yotter and Wilson, 2003). Over the last decade, the use of color image sensors for chemical quantification has been increasing very rapidly. These include color flatbed scanners (Lavigne et al., 1998; Taton et al., 2000; Rakow and Suslick, 2000) and digital color cameras (Jenison et al., 2001; Filippini et al., 2003; Abe et al., 2008; Martinez et al., 2008; Stich et al., 2009). Figure 2.1 shows the conceptual diagrams of the color-sensing and color emission devices utilizing thin-film color filter arrays composed of three primary colors (red, green, and blue). The photodetector array in the color image sensor records the incident photon intensity spectrally separated according to the three different colors of on-chip Bayer filters, as in Figure 2.2a. Since a wide

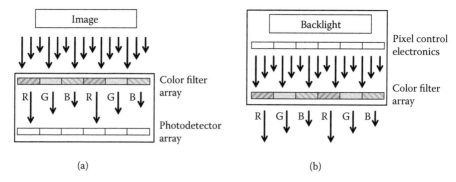

(a)   (b)

FIGURE 2.1 Simplified cross-sectional views of color image sensor (a) and color display devices (b). Arrays of thin-film color filters (red, green, blue), originally intended for sensory purposes, provide spectral sensitivities for analytical applications.

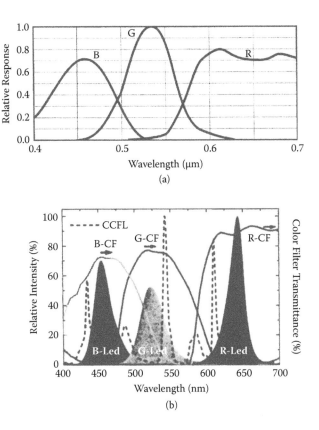

FIGURE 2.2 (a) Spectral ranges of three primary color filters of typical image sensors. (Reprinted from Holst, G. C., and T. S. Lomheim, *CMOS/CCD Sensors and Camera Systems* (2nd ed.), Bellingham, WA: SPIE Press, 2011, p. 143. With permission.) (b) Emission spectra of backlights (cold cathode fluorescent lamp and light-emitting diode) and transmission ranges of three color filters from typical liquid crystal display screens. (Reprinted from Lee, J.-H., D. N. Liu, and S.-T. Wu, *Introduction to Flat Panel Displays*, United Kingdom: John Wiley & Sons, 2008, p. 170. Copyright © 2008 John Wiley & Sons. With permission.)

variety of luminophores emit lights in the visible range, this device can serve as the analytical instrument for colorimetric chemical quantification.

Another innovation in recent photonics development is the advent of optoelectronic display devices, including LCD and DLP devices. Similarly, the ability to emit light in a selected range with the built-in color filters, as shown in Figure 2.2b, makes it very attractive as a ubiquitous light source for chemical quantification. This type of emission device can be utilized as the illumination (for transmittance or reflection measurements) and excitation (for fluorescence emission) light sources required for optical interrogation (Batchelor and Jones, 1998; Filippini and Lundstrom, 2002; Filippini et al., 2003). We envision that the color imaging devices will be key components to develop low-cost, portable sensors for environmental and industrial monitoring and point-of-care testers in resource-poor settings.

2.1.2 Color Space

There are several color space models to quantitatively describe the color intensity with respect to the color-matching function established by the International Commission on Illumination (CIE). The representative color systems are RGB (red, green, blue), CMYK (cyan, magenta, yellow, black), CIELAB/CIELUV (lightness, color differences), and HSL (hue, saturation, lightness). Each system has its own characteristic advantages in describing the colors quantitatively, and thus is complementary to other ones. Both the RGB and CMYK systems are rather device dependent compared to the HSL and CIE systems that are designed to be perceptually uniform to approximate human vision. Some recent reports on colorimetric chemical quantification show that the CIELAB or HSL system is useful to express a variation in brightness of one color (i.e., a single emission peak), while the CMYK or RGB system appears to be more suitable in the case of active color changes (i.e., ratiometric change) (Abe et al., 2008; Martinez et al., 2008; Stich et al., 2009).

Despite a lossy compression format, the Joint Photographic Experts Group (JPEG) file format with the Exchangeable Image File Format (EXIF) algorithm can be commonly used for storing the color information because it is supported by most imaging devices and image software. Other lossless formats such as RAW or BMP are also available, but with larger file sizes. All data can be processed by one of these formats with 8-bit per single color information (i.e., 24-bit RGB with 256-level per each color).

2.2 PROSPECTIVE LUMINOPHORES FOR COLORIMETRIC DETERMINATION

Luminescence is an emission of light resulting from electronically excited molecules via chemical reaction, light, or other stimuli, including electricity, temperature, sound, and pressure. The molecule showing the luminescence is called a luminophore. In this chapter, we focus on the sensing based on photoluminescence, i.e., luminescence induced by light. Furthermore, photoluminescence can be divided into fluorescence and phosphorescence depending on the electron spin states involved during the photon emission process.

Luminescence can be used for sensing purposes in two different ways. It may directly respond to the concentration of target analytes by showing a change in the emission intensity or in the Stokes shift. Alternatively, a luminophore can be attached to a reporter molecule that can specifically interact with a target analyte. Luminophores can originate from various sources, such as organic dyes, polymer conjugates, semiconductor quantum dots, Au and Ag nanoparticles, carbon nanomaterials, fluorescent proteins, and metal complexes. The peak absorption and emission wavelengths of these materials vary and can even be tuned chemically or physically. Since the wavelength of the light source and detector in our colorimetric sensor platform is currently limited to the visible range, this chapter focuses on luminescent materials showing peak excitation and emission wavelengths in the visible range.

2.2.1 ORGANIC DYES

Organic dyes are the most widely used reporters in fluorescence sensing because of their economical cost and easy modification of their structures, which can diversify spectral properties. In addition, the derivative forms of a fluorescence dye can be used to label proteins and other biological samples. Organic fluorescent dyes that can be used in our sensor platform are shown in Table 2.1. One of the important selection criteria is that the dye should show a large Stokes shift (>50 nm) to reduce the interference of excitation light in sensing fluorescence emission. However, fluorescence dyes showing a small Stokes shift (less than 50 nm) may also be used in our sensor. For example, the peak excitation and emission wavelengths of fluorescein are 488 and 514 nm, respectively. The blue-filtered light from an LCD source can excite the molecule, but the selection of green in the CCD sensor cannot completely eliminate the excitation light, thus resulting in a poor signal-to-background ratio. In this case, a proper cutoff filter can be introduced in front of the image sensor.

TABLE 2.1
Peak Excitation and Emission Wavelengths of Various Organic Luminophores Potentially Suitable for Colorimetric Determination

Luminophore	Excitation Wavelength (nm)	Emission Wavelength (nm)
Acridine yellow	470	550
Acriflavin	436	520
7-Aminoactinomycin D-AAD	546	647
Astrazon Orange R	470	540
Auramine	460	550
Aurophosphine G	450	580
Berberine sulfate	430	550
Brilliant Sulpho Flavin FF	430	520
Coriphosphine O	460	575
DiA	456	590
Fura Red	472(low $[Ca^{2+}]$), 436(high $[Ca^{2+}]$)	657(low $[Ca^{2+}]$), 637(high $[Ca^{2+}]$)
Genacryl Pink 3G	470	583
Mithramycin	450	570
NBD	465	535
NBD amine	450	530
Phosphine 3R	465	565
Pontochrome Blue Black	535–553	605
Procion yellow	470	600
Rhodamine 5 GLD	470	565
Rhodamine B	540	625
Rhodamine B 200	523–557	595
Rhodamine B Extra	550	605
Sevron Orange	440	530
Sulpho Rhodamine G Extra	470	570

The applications of organic dye for sensing are diverse. Some fluorescence dyes respond to a change in the environment, such as solvent polarity, pH, intermolecular interaction, or electric field. Generally, organic dye molecules are easily photobleached. Their photostability can be improved by incorporating them into polymer (Landfester, 2006) or silica (Gerion et al., 2001) particles. The organic dyes encapsulated in these particles exhibit higher fluorescence quantum yield and are less prone to photobleaching than free organic dyes (Yao et al., 2006).

2.2.2 NANOPARTICLES

Quantum dots (QDs) are typically 2–60 nm diameter nanocrystals made from Group II/VI and III/V semiconductors. Examples are CdSe, CdTe, InP, and InGaP, which are all commercially available now. In general, quantum dots can absorb light ranging from ultraviolet (UV) to near infrared (IR), but their absorption decreases as the incident light wavelength increases. The emission bands of quantum dots are very narrow and symmetric. This makes quantum dots more attractive than the organic dye because of easy separation of the emission photons from the excitation photons. Quantum dots can be efficiently excited with the blue-filtered light from an LCD light source. The emission light can be observed using green or red selection in a CCD camera. Particularly, red quantum dots will show better signal-to-background ratios than green or yellow-orange quantum dots because of less overlap between the excitation and emission bands. Since quantum dots for bioconjugation are commercially available, constructing a quantum dot biosensor is possible using our sensor platform. Quantum dots show strong brightness because of their high molar absorptivity ($\sim10^6$ M^{-1} cm^{-1}) and quantum yield, which cannot be easily achieved with organic dyes. In addition, unlike organic dyes, quantum dots have shown excellent chemical and photostability. While quantum dots have gained popularity for labeling or tagging biological samples, the toxicity of Cd-based quantum dots has been a major concern.

Recently, porous silicon and silicon nanoparticles have received attention because they may replace Cd-containing quantum dots. Silicon nanoparticles show a broad luminescence spectrum in the visible range. The peak emission wavelength shifts depending on the excitation wavelength, because the silicon particles have a size distribution (Veinot, 2006). Therefore, the silicon nanoparticles can be excited with blue-filter light and detected using the red option in the CCD detector.

Nanoparticles consisting of noble metal atoms, such as Au and Ag, absorb light in a visible range in a size-dependent manner. Unlike semiconductor quantum dots, the light absorption and emission of noble metal nanoparticles are related to plasmons, a collective oscillation of free electrons upon the incidence of light. Au nanoparticle solution scatters and absorbs light efficiently but exhibits very weak fluorescence; therefore, the absorption property of Au nanoparticles is mainly used for sensing purposes. In addition, since Au acts as an efficient fluorescence quencher, the Au–fluorescence dye conjugate can be used for various sensing applications using the de-quenching effect.

2.2.3 LUMINESCENT METAL COMPLEXES

The metal complexes of lanthanides and noble metal ions exhibit long lifetime luminescence. Among the lanthanide ions, Tb^{3+} and Eu^{3+} exhibit relatively high emission intensities in the visible range. Lanthanide ions themselves do not absorb photons very well, and so they are not good photon emitters. But, the formation of a complex with heterocyclic chelating agent enhances the luminescence. The aromatic hetero-cycle chelating groups, called antennas, primarily absorb photon energy and then relay the excitation energy to the lanthanide. This energy transfer can increase the molar absorptivity to above 10,000 M^{-1} cm^{-1}, which is sufficient for luminescence sensing. Unfortunately, these metal complexes cannot be employed in our sensor because of their weak absorption in the 400–600 nm range.

Ru^{2+}, Os^{2+}, and Re^{2+} ions can form different types of metal-ligand complexes showing long-lifetime emission. A representative example is the Ru(bpy)$_3$$^{2+}$ com-plex formed between Ru^{2+} and the tris-(2,2'-bipyridine) ligand (Balzani et al., 2001). Its absorption spectrum is very broad, showing a peak absorbance around 470 nm (molar absorptivity: 10,000–30,000 M^{-1} cm^{-1}). A broad emission band is observed ranging from 600 to 670 nm. Thus, these compounds have the spectral characteristic required for our sensor platform. Its origin of photon emission is different from that of the lanthanide complexes. It is phosphorescence resulting from the metal-ligand charge transfer (MLCT). The lifetime of this photon emission is much longer than that of the fluorescence emission. The emission is highly sensitive to oxygen concen-tration and has been used in our sensors, which will be discussed in the next section.

A porphyrin complex containing Pt or Pd is another example of a phosphorescent metal complex. A porphyrin structure can be easily found in nature from hemoglo-bin or chlorophyll. The porphyrin complex absorbs light at narrow-band regions at 360–400 nm and 500–550 nm and emits red phosphorescence having a long life-time (0.01 to 1 ms) (Papkovsky et al., 2000). These porphyrin complexes can be applied in our sensor. They have been used for sensing oxygen, glucose, and lactate concentrations.

2.2.4 CONJUGATE POLYMERS

Conjugate polymers are the compounds with alternating single and double bonds (or aromatic units) along the polymer chain. The polarizable π-electrons extended along the conjugated backbone will determine the optical properties of conjugate polymers. They have high molar absorptivity, reaching 10^6 M^{-1} cm^{-1}, and display strong fluorescence. The conjugate polymer can be prepared as a thin film and used as a sensor relying on super quenching or super enhancement effect (Thomas et al., 2007). The quenching in conjugate polymer arises from the termination of exciton migration and occurs in a collective manner. One quencher molecule can simultane-ously quench a large number of fluorescent monomeric units, and thus it is called super quenching. On the other hand, the super enhancement effect occurs when the quencher is removed from the conjugate polymer. The change in chain confor-mation of the conjugate polymer modifies the spectral properties of the polymer, i.e., absorption and emission. Examples of conjugate polymers are polyacetylene, polythiophene, polypyrrole, polyaniline, and polyparaphenylene vinylene. A rational design of the conjugate polymer can provide fine-tuning for its excitation and emis-sion wavelengths that fit the colorimetric sensing modality.

2.2.5 FLUORESCENT PROTEINS

The green fluorescent protein (GFP) is a naturally fluorescent protein that was first extracted from a jellyfish species called *Aequorea victoria*. It is an exceptional protein that can be stably expressed as a fusion protein in a species other than the jellyfish. The wild-type GFP has two excitation peaks at 395 and 470 nm and the emission peak is at 509 nm. The GFP variants have revealed a shift in excitation peaks while maintaining emission peak positions similar to that of the wild-type GFP (Tsien, 1998). So the over-all excitation and emission spectra of the GFP variant are similar to those of fluorescein dye. This makes the GFP variant useful for fluorescence imaging. Furthermore, a different color of fluorescent protein was derived from the GFP mutant variants. Blue, red, cyan, and yellow fluorescent proteins are possible (Shaner et al., 2004). Recently, other GFP-like fluorescent proteins were derived from sources other than the jellyfish, which will allow diverse spectral characteristics for fluorescent proteins. GFP and its variants have served as unique tools in live cell imaging for protein tagging, gene expression monitoring, and drug and genetic screens.

2.3 COLOR CAMERA AS PHOTODETECTOR FOR DISSOLVED OXYGEN QUANTIFICATION

2.3.1 OPTICAL OXYGEN SENSING

Optical oxygen sensors are becoming dominant over electrochemical types, because they are nondestructive sensors that do not consume oxygen, and thus do not perturb the oxygen environment during the measurement. Other advantages include (1) easy and simple device miniaturization, (2) capability of remote sensing in a noninvasive and noncontact mode, and (3) capability of two-dimensional imaging of oxygen distribution. Optical oxygen sensors use the mechanism of oxygen quenching as shown in Figure 2.3a. Quenching is the phenomenon of the emission intensity decrease of a

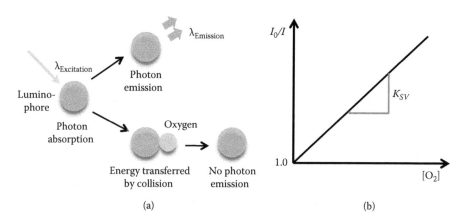

(a)   (b)

FIGURE 2.3   (a) Principle of luminescence quenching by molecular oxygen (λ_{ex}, excitation wavelength; λ_{em}, emission wavelength) depicting the luminescence process in the absence of oxygen and the deactivation of luminophore by oxygen. (b) Stern-Volmer plot based on Equation 2.1.

luminophore in the presence of a quencher. In this case, an oxygen molecule acts as a powerful quencher of the electronically excited state of the luminophore. Oxygen-sensitive luminophores include metal complex organic dyes, dual emitters, and fullerenes (Amao, 2003; Wang et al., 2010a). Recent studies have also shown the use of ruthenium and porphyrin complexes as luminophores for oxygen sensors. These agents are typically immobilized within oxygen-permeable matrices for implementing a sensor device, and their spectral properties are not significantly altered by the immobilization. Oxygen concentration can be determined from the Stern-Volmer relationship by measuring intensity, lifetime, or phase shift as follows (Mills, 1997):

$$I_0/I \; (= \tau_0/\tau = \Phi_0/\Phi) = 1 + K_{SV} \, [O_2] \qquad (2.1)$$

where I_0 and I (τ, Φ) represent the steady-state luminescence intensities without and with the quencher (here oxygen). Alternately, τ and Φ represent the lifetime (decay time) of luminescence and the phase angle shift (time delay) of luminescence from a sinusoidal excitation light, respectively. K_{sv} is the Stern-Volmer quenching constant, and $[O_2]$ is the oxygen concentration. Therefore, for an ideal condition, the ratio of luminescence intensities without and with oxygen (I_0/I) becomes linearly proportional to the oxygen concentration as plotted in Figure 2.3b.

Several colorimetric oxygen-sensing methods based on absorption or emission measurements have been reported (Eaton, 2002; Evans and Douglas, 2006; Evans et al., 2006). Although the results were semiquantitative, these approaches pioneered the concept of colorimetric oxygen quantification. Recently, several research groups, including us, have reported the use of color cameras for quantitatively determining the oxygen concentration. These intensity-based oxygen sensing has demonstrated the analytical capability of color imaging devices (Thomas et al., 2009; Park et al., 2010; Park, 2011; Wang et al., 2010b; Shen et al., 2011). We used a ruthenium complex as the oxygen-sensing luminophore. This shows a great Stokes shift by emitting orange light (about 590 nm peak) when excited with blue light (about 470 nm peak). The complex is ideal for colorimetric oxygen quantification. This section summarizes how a color CCD camera and simple color analysis can serve as reliable analytical instruments.

2.3.2 COMMERCIAL SOL-GEL SENSOR

An oxygen sensor patch (RedEye™, RE-FOX-8, 8 mm diameter, OceanOptics), which is commercially available, was used for a proof-of-concept demonstration. This oxygen sensor patch contained ruthenium complex immobilized within a sol-gel matrix. A color camera and a spectrophotometer were used to measure the oxygen concentration, and the two results were compared under the same conditions. A color camera (DS-5M, Nikon) that employs a CCD device (ICX282AQ, 5 megapixels, Bayer-masked, Sony) was used for image capturing. A spectrofluorometer was used as the detector. A blue light-emitting diode (LED) was used as the excitation source (peak wavelength 470 nm) for both methods. A long-wave pass filter (cut-on wavelength 500 nm) was used to minimize the intense blue excitation wavelengths that can interfere with the sensitivity of red pixel photodetectors.

Figure 2.4a shows the emission spectra of the ruthenium complex in the 550–700 nm range in response to the oxygen concentration. The data clearly demonstrate that the peak emission intensity at 595 nm decreases as the oxygen concentration increases. This oxygen-responsive emission range is almost identical to the spectral range of the red-responsive pixels of the color imager, as in Figure 2.2a. An exemplary image analysis result is shown in Figure 2.4b that includes a red-extracted image of the sensor patch at 0% oxygen solution. The Stern-Volmer plots (Equation 2.1) in Figure 2.5 include the results of various image analysis and spectrometric data. The spectrometry data represent the intensities at the peak wavelength (595 nm) of Figure 2.4a. The image analysis data represent the mean values of digital color intensities (between 0 and 255) from the histograms. The red color intensity data obtained by the CCD camera show a broader linearity and better sensitivity than the integrated fluorescence intensity data collected by the conventional

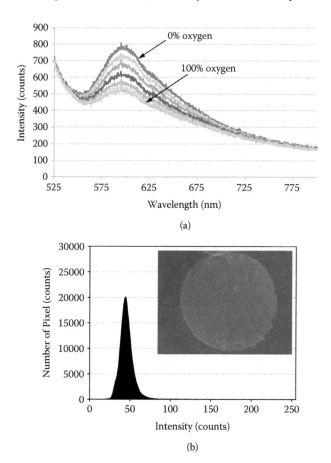

(a)

(b)

FIGURE 2.4 (a) The emission spectra of a commercial oxygen-sensitive patch in various dissolved oxygen concentrations. (b) Red-extracted images of the RedEye patch (8 mm diameter) and its histogram of red color intensity. (Reprinted from Park, J., W. Hong, and C.-S. Kim, *IEEE Sensors Journal*, 10(12), 1855–1861, 2010. Copyright © 2010 IEEE. With permission.)

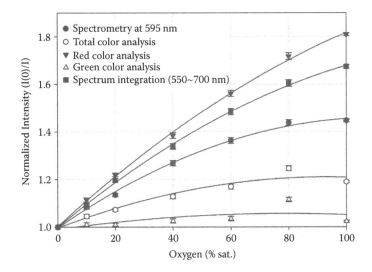

FIGURE 2.5 Stern-Volmer plots of various methods: (from the top) red intensity of image (▲), spectrum areal integration of spectrometry (Ж), emission peak intensity of spectrometry (♦), total color intensity of image (■), and green intensity of image (x). (Reprinted from Park, J., W. Hong, and C.-S. Kim, *IEEE Sensors Journal*, 10(12), 1855–1861, 2010. Copyright © 2010 IEEE. With permission.)

spectrometer (Figure 2.5). Only the red component from the original RGB color image was extracted by ImageJ, a free image processing software developed by the National Institutes of Health (Abramoff et al., 2004). This procedure serves as a virtual band-pass filter by removing unnecessary "noise" colors (i.e., green-blue range) to effectively "mine" only the oxygen-related information.

2.3.3 OPTOFLUIDIC HYDROGEL SENSOR

We have extended our previous study by using lab-made sensor assemblies. Ruthenium complex, dichlorotris(1,10-phenanthroline) ruthenium (II) hydrate was embedded in photopatterned polyethylene glycol (PEG) hydrogels to use as the sensing element. The fabrication steps for the array are illustrated in Figure 2.6a. The hydrogel precursor solution was injected into an assembly that consisted of two glass slides (1 × 3 in.) and a spacer. One slide was surface treated with a silanization agent to promote the adhesion between the glass and PEG. A scotch tape film was adhered on the other slide to prevent the PEG layer from being attached to the slide. After the PEG layer was photopatterned by illuminating UV light through a photomask and rinsing unreacted residue, a silicone CoverWell™ sheet (32 × 19 × 0.5 mm) was placed to cover the hydrogels and form a sealed optofluidic assembly. Twenty circular sensors (0.5 mm diameter, 120 µm thick) were patterned within a 1 cm² area.

Figure 2.6b is the red-extracted images of two-dimensional oxygen sensors in 0, 20, and 100% oxygen-saturated water. It can be recognized that the red intensity decreases with increasing dissolved oxygen concentration due to the quenching reaction. The absolute red intensity of each sensing spot showed variations for

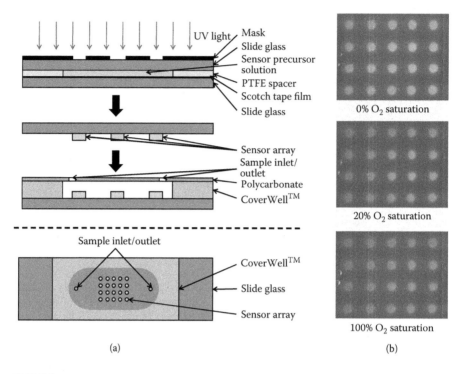

FIGURE 2.6 (See color insert.) Optofluidic dissolved oxygen sensor assembly. (a) Photo-patterning of PEG hydrogel array and layout of the assembled system. (b) Red-extracted images of PEG array in 0, 20, and 100% oxygen saturated water. (Reprinted from Park, J., W. Hong, and C.-S. Kim, *IEEE Sensors Journal*, 10(12), 1855–1861, 2010. Copyright © 2010 IEEE. With permission.)

three reasons: inhomogeneity in excitation illumination intensity and variations of the sensor thickness and the amount of ruthenium complex in the sensors. However, the calibration curve for each sensor can be constructed by comparing the intensity of each sensor in the image, which is the inherent advantage of the two-dimensional image analysis. Figure 2.7 shows the Stern-Volmer plot of the average of the red intensities of 20 sensors and that obtained with the spectrometer. The red intensity of each sensor was obtained by averaging the red intensity over the sensor area (0.5 mm diameter). Similar to the above, the red analysis data showed better sensitivity and a broader linear range than the spectrometric analysis data.

2.4 NONTRADITIONAL EMISSION DEVICES AS LIGHT SOURCE FOR OXYGEN QUANTIFICATION

Light-emitting diode or traditional bulky, broad-band lamps with filters can be commonly used to excite luminophores. A white LED or an LCD screen has been employed for determining gaseous oxygen both quantitatively and qualitatively. The same color camera setup was successfully used to characterize red emission from sensor films based on colorimetric intensity measurements.

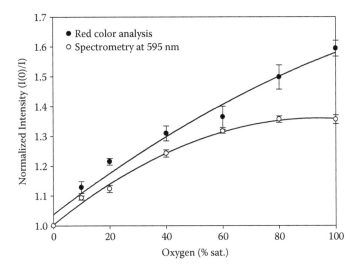

FIGURE 2.7 Stern-Volmer plots of the PEG oxygen sensor array with respect to dissolved oxygen based on spectrum and red color analysis. (Reprinted from Park, J., W. Hong, and C.-S. Kim, *IEEE Sensors Journal*, 10(12), 1855–1861, 2010. Copyright © 2010 IEEE. With permission.)

2.4.1 WHITE LED AS EXCITATION SOURCE

We further explored the possibility of using a filter-free white light source for quantitative imaging of gaseous oxygen. The principal method of implementing the white LED is to add a phosphor material in the blue LED. This implies that the blue wavelength range is a major emission peak of the white LED that can excite the oxygen-sensitive ruthenium complex. As shown in Figure 2.8a, a broad-band white LED (LS-450 with LED-WHITE, OceanOptics) was used for quantifying gaseous oxygen. The same commercial oxygen-sensitive patch as above was used to examine this new approach. Gas samples of various oxygen percentages were prepared with mass flow controllers. The sample gas was uniformly delivered over the sensor surface through a glass tube (1 mm inner diameter) at a moderate flow rate of 2 liter/min.

The spectral output of the white LED has a peak emission around 475 nm that matches well with the excitation wavelength of the ruthenium complex. A simple red color analysis of CCD images showed that the white LED without any filter performed similarly to the blue LED with a filter. Figure 2.8b shows a graph comparing the normalized red color sensitivities of three different optical configurations. Although the white LED configuration without filtering shows the least linearity over the entire 0–100% oxygen range, it has the highest Stern-Volmer (I_0/I) sensitivity in the 0–20% oxygen range. This concentration range is used for various medical and biochemical process monitoring.

For an oxygen distribution mapping, only part of the sensor patch was directly exposed to the oxygen flow to generate an oxygen gradient over the patch surface. The right-side image in Figure 2.9a was the result of subtracting a red fluorescence intensity image in the presence of the oxygen flow (middle) from the reference image obtained

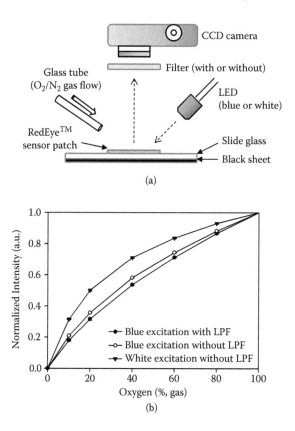

(a)

(b)

FIGURE 2.8 (a) The sensor imaging setup with a color CCD camera for gaseous oxygen quantification. A long-wave pass filter, blue LED, and white LED were selectively used for a series of measurements. The microscope setup is not shown. (b) Normalized Stern-Volmer plots to compare the performance of the three different imaging configurations. (Reprinted from Park, J., and C.-S. Kim, *Sensor Letters*, 9(1), 118–123, 2011. Copyright © 2011 American Scientific Publishers. With permission.)

with no oxygen flow (left side). Figure 2.9b shows the two-dimensional intensity gradient in the selected region of the subtracted image in Figure 2.9a (dashed line). We compared two measurement configurations: standard blue excitation with filter and broad-band white illumination without filter. Normalized intensity profiles of the two configurations along the dashed line showed good agreement between the two results (Figure 2.9c). We anticipate that the broad-band white excitation configuration can be used for simultaneous quantification of multiple target analytes. Several luminophores that have different excitation and emission wavelengths can be analyzed by using the white light source, digital color imagers, and color analysis methods.

2.4.2 LCD MONITOR AS EXCITATION SOURCE

The traditional excitation light sources, such as diodes and lamps, are rather limited to one-dimensional illumination over sample surfaces. A uniform illumination of

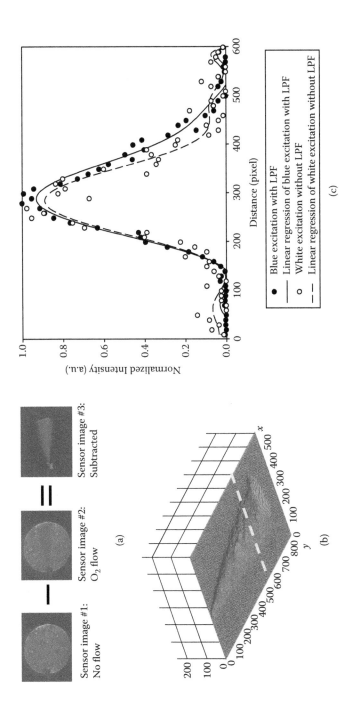

FIGURE 2.9 Mapping of oxygen gradient on sensor surface (8 mm diameter) created with a capillary tube. (a) Subtraction of two images (with oxygen flow and without oxygen flow) to eliminate background. (b) Plot of the differential intensity profile of the region of interest (ROI) in the subtracted image in (a). (c) Normalized intensity profiles along the dashed line in (b) (-●-, blue LED excitation with filter; -○-, white LED excitation without filter). (Reprinted from Park, J., and C.-S. Kim, *Sensor Letters*, 9(1), 118–123, 2011. Copyright © 2011 American Scientific Publishers. With permission.)

excitation light over a target area is important for analyzing the spatial distribution of chemicals. Two-dimensional display screens can be utilized for this application. Their emission colors (i.e., wavelength ranges) and intensity can be easily controlled by a computer. Commercial LCD monitors mix three primary colors (red, green, and blue) to display true color (i.e., 16,777,216 different colors with 24-bit RGB color space). In principle, three major wavelength ranges from their backlight (usually the broad-band cold cathode fluorescent lamp (CCFL)) can be selected with controlled intensities. Several new techniques have been reported where these monitors are applied for analytical purposes. In particular, Filippini and coworkers (Filippini and Lundstrom, 2002; Filippini et al., 2003) implemented the idea of the computer screen photo-assisted technique (CSPT), where commercial monitors were used as an illumination light source for analytical applications. These examples include illuminating samples for transmission/absorption measurements and exciting luminescent samples for emission measurements.

For oxygen quantification, we used an LCD monitor (HP 2159m, 21.5 in. color LCD monitor, Hewlett-Packard) as the excitation light source. Mesoscale test platforms (8×8 cm^2) incorporating a fluidic channel and a sensor coating were prepared and attached on an LCD screen, as shown in Figure 2.10a. Planar sensor films formulated in our laboratory were used for oxygen imaging. This was prepared by mixing silica particles containing ruthenium complex with silicone prepolymer, and spin coating and curing the mixture on glass plates. Blue light (470 nm) from the LCD screen was used to excite the ruthenium complex films that respond to gaseous oxygen. The color camera provided significant advantages for storing colorimetric digital data with two-dimensional information and analyzing its color change and gradient over a captured image area.

Figure 2.10b shows a reconstructed image of the spatial distribution of the relative intensity (I_0/I) of the Stern-Volmer relation. Nitrogen gas was initially injected into two channel branches to obtain a reference image (equivalent to I_0), followed by switching the gas for the lower branch to air (21% oxygen) for a sample image (equivalent to I). The reference image was divided by the second image on a pixel-by-pixel basis to obtain the final Stern-Volmer image. Figure 2.10c shows the oxygen concentration profiles obtained from four different locations of the fluidic channel as indicated by dotted lines in Figure 2.10b. Each profile represents the oxygen concentration gradient along the dotted lines. The difference of oxygen content between the upper and lower branches was clearly observed (V1 location). The oxygen concentration gradient becomes less steep along the downstream (from the right to the left in Figure 2.10b), because the gases introduced from the two channels mix together in the main channel. These results show that the combination of an LCD and a color camera enables a uniform illumination over a large area to image spatial distribution of chemicals. It is anticipated that the quantification of multiple target analytes is possible with variable wavelength ranges emitted from the LCD monitor. Furthermore, time-resolved imaging will also be possible with an application-specific camera paired with a synchronized display device with sinusoidal or pulsatile emission.

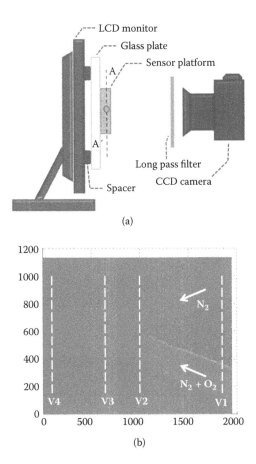

(a)

(b)

FIGURE 2.10 (See color insert.) (a) Measurement setup with an LCD monitor as the excitation light source and a color camera as the photodetector. A color camera takes pictures of the fluidic sensor platform (8×8 cm^2) installed in close proximity to an LCD screen that provides an excitation blue light (470 nm) with uniform intensity over the sensor coating. (After Park, S., S. G. Achanta, and C.-S Kim, Fluorescence Intensity Measurements with Display Screen as Excitation Source, *Progress in Biomedical Optics and Imaging, Proceedings of SPIE*, Orlando, FL, April 25–29, 2011, Paper 802509. Permission required from SPIE.) (b) Stern-Volmer image of oxygen distribution (equivalent to I_0/I). (c) Oxygen profiles at various locations defined in (b) (V1, V2, V3, and V4), showing a nitrogen and 20% oxygen fluxes at upper and lower branches, respectively. *(continued)*

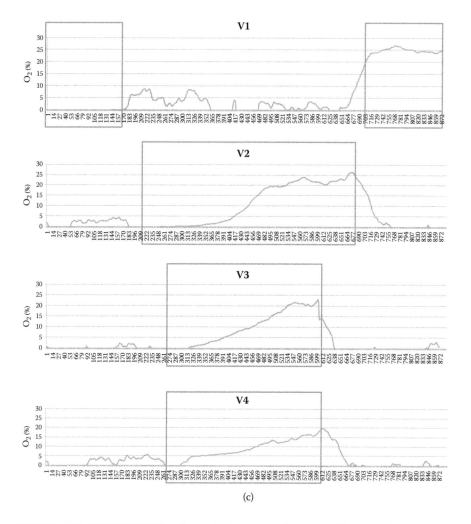

(c)

FIGURE 2.10 (CONTINUED) (See color insert.) (c) Oxygen profiles at various locations defined in (b) (V1, V2, V3, and V4), showing a nitrogen and 20% oxygen fluxes at upper and lower branches, respectively.

2.5 CONCLUSION

A wide variety of luminophores from organic dyes, inorganic nanoparticles, polymers, and proteins are available that exhibit large Stokes shifts within visible ranges. Therefore, these agents are potentially amenable for colorimetric determination based on this approach of adopting easily accessible colorimetric devices as analytical instruments.

As an exemplary proof-of-concept demonstration, rather nontraditional optoelectronic devices in analytical sciences were successfully used for oxygen determination. The ruthenium complex luminophores embedded in sensor matrices were excited by the blue wavelength range (about 470 nm peak) emitted from the white LED and the LCD monitor. The color camera exhibited good sensitivity and linearity to gaseous and dissolved oxygen samples. Both qualitative and quantitative analyses were possible with relatively simple colorimetric image analysis. The combination of LCD and color camera especially enables a uniform illumination over a large area to image spatial distribution of chemicals and to analyze multiple target analytes simultaneously.

In general, the CCD cameras, especially the cooled monochromatic ones, are highly sensitive and low-noise imaging devices compared with their CMOS counterpart, making them more suitable for high-quality imaging in scientific research. The advent of submicron-scale memory chip technology, however, allowed the performance of CMOS camera chips to become more competitive than that of the CCD. The CMOS chips are fabricated by economical standard integrated circuit (IC) processes without tailoring their process sequences. More importantly, they consume less power, which is essential for portable applications. In fact, the majority of portable imaging devices in mobile cellular phones and webcams in the market are economical CMOS devices. Furthermore, novel emission devices based on new materials are emerging to provide new functionalities. One representative example is the organic light-emitting diodes (OLEDs) that can be seamlessly integrated in flexible platforms. Much preceding research relied on commercially available products. However, customized optoelectronic components can be developed to obtain suitable emission intensity, photonic sensitivity, spectral responsivity and dynamic range, etc. These innovations in optoelectronics areas will largely expedite the progress of economic analytical instruments based on colorimetric interrogation.

REFERENCES

Abe, K., K. Suzuki, D. Citterio, Inkjet-printed microfluidic multianalyte chemical sensing paper, *Analytical Chemistry*, 80, 6928–6934, 2008.

Abramoff, M. D., P. J. Magelhaes, S. J. Ram, Image processing with ImageJ, *Biophotonics International*, 11, 36–42, 2004.

Amao, Y., Probes and polymers for optical sensing of oxygen, *Microchimica Acta*, 143(1), 1–12, 2003.

Babin, S., A. Bugrov, S. Cabrini, S. Dhuey, A. Goltsov, I. Ivonin, E.-B. Kley, C. Peroz, H. Schmidt, V. Yankov, Digital optical spectrometer-on-chip, *Applied Physics Letters*, 95(4), 041105, 2009.

Balzani, V., P. Ceroni, A. Juris, M. Venturi, S. Campagna, F. Puntoriero, S. Serroni, Dendrimers based on photoactive metal complexes: Recent advances, *Coordination Chemistry Reviews*, 219–221, 545–572, 2001.

Batchelor, J. D., B. T. Jones, Development of a digital micromirror spectrometer for analytical atomic spectroscopy, *Analytical Chemistry*, 70, 4907–4914, 1998.

Eaton, K., A novel colorimetric oxygen sensor: Dye redox chemistry in a thin polymer film, *Sensors and Actuators B: Chemical*, 85, 42–51, 2002.

Evans, R. C., P. Douglas, Controlling the color space response of colorimetric luminescence oxygen sensors, *Analytical Chemistry*, 78, 5645–5652, 2006.

Evans, R. C., P. Douglas, J. A. G. Williams, D. L. Rochester, A novel luminescence-based colorimetric oxygen sensor with a "traffic light" response, *Journal of Fluorescence*, 15, 201–206, 2006.

Filippini, D., I. Lundstrom, Chemical imaging by a computer screen aided scanning light pulse technique, *Applied Physics Letter*, 81(20), 3891–3893, 2002.

Filippini, D., S. P. S. Svensson, I. Lundstrom, Computer screen as a programmable light source for visible absorption characterization of (bio)chemical assays, *Chemical Communication*, 9, 240–241, 2003.

Gerion, D., F. Pinaud, S. C. Williams, W. J. Parak, D. Zanchet, S. Weiss, A. P. Alivisatos, Synthesis and properties of biocompatible water-soluble silica-coated CdSe/ZnS semi-conductor quantum dots, *Journal of Physical Chemistry B*, 105(37), 8861–8871, 2001.

Holst, G. C., T. S. Lomheim, *CMOS/CCD sensors and camera systems* (2nd ed.), p. 143, London: Nature Publishing Group, SPIE Press, 2011.

Jenison, R., S. Yang, A. Haeberli, B. Polisky, Interference-based detection of nucleic acid targets on optically coated silicon, *Nature Biotechnology*, 19, 62–65, 2001.

Landfester, K., Synthesis of colloidal particles in miniemulsions, *Annual Review of Materials Research*, 36, 231–279, 2006.

Lavigne, J. J., S. Savoy, M. B. Clevenger, J. E. Ritchie, B. McDoniel, S.-J. Yoo, E. V. Anslyn, J. T. McDevitt, J. B. Shear, D. Neikirtk, Solution-based analysis of multiple analytes by a sensor array: Toward the development of an "electronic tongue," *Journal of the American Chemical Society*, 120, 6429–6430, 1998.

Lee, J.-H., D. N. Liu, S.-T. Wu, *Introduction to flat panel displays*, p. 170, United Kingdom: John Wiley & Sons, 2008.

Martinez, A. M., S. T. Philips, E. Carrilho, S. W. Thomas, H. Sindi, G. M. Whitesides, Simple telemedicine for developing regions: Camera phones and paper-based microfluidic devices for real-time, off-site diagnosis, *Analytical Chemistry*, 80, 3699–3707, 2008.

Mills, A., Optical oxygen sensors, *Platinum Metals Review*, 41(3), 115–127, 1997.

Papkovsky, D. B., T. O'Riordan, A. Soini, Phosphorescent porphyrin probes in biosensors and sensitive bioassays, *Biochemical Society Transactions*, 28, 74–77, 2000.

Park, J., W. Hong, C.-S. Kim, Color intensity method for hydrogel optical sensor array, *IEEE Sensors Journal*, 10(12), 1855–1861, 2010.

Park, J., C.-S. Kim, A simple oxygen sensor imaging method with white light-emitting diode and color charge-coupled device camera, *Sensor Letters*, 9(1), 118–123, 2011.

Park, S., S. G. Achanta, C.-S Kim, Fluorescence intensity measurements with display screen as excitation source, *Progress in Biomedical Optics and Imaging, Proceedings of SPIE*, Orlando, FL, April 25–29, 2011, Paper 802509.

Rakow, N. A., K. S. Suslick, A colorimetric sensor array for odour visualization, *Nature*, 406, 710–713, 2000.

Shaner, N. C., R. E. Campbell, P. A. Steinbach, B. N. G. Giepmans, A. E. Palmer, R. Y. Tsien, Improved monomeric red, orange and yellow fluorescent proteins derived from *Discosoma* sp red fluorescent protein, *Nature Biotechnology*, 22, 1567–1572, 2004.

Shen, L., M. Ratterman, D. Klotzkin, I. Papautsky, Use of a low-cost CMOS detector and cross-polarization signal isolation for oxygen sensing, *IEEE Sensors Journal*, 11(6), 1359–1360, 2011.

Stich, M. I. J., S. M. Borisov, U. Henne, M. Schaferling, Read-out of multiple optical chemical sensors by means of digital color cameras, *Sensors and Actuators B*, 139, 204–207, 2009.

Syms, R. R. A., Advances in microfabricated mass spectrometers, *Analytical and Bioanalytical Chemistry*, 393(2), 427–429, 2009.

Taton, T. A., C. A. Mirkin, R. L. Letsinger, Scanometric DNA array detection with nanoparticle probes, *Science*, 289, 1757–1760, 2000.

Thomas, P. C., M. Halter, A. Tona, S. R. Raghavan, A. L. Plant, S. P. Forry, A noninvasive thin film sensor for monitoring oxygen tension during in vitro cell culture, *Analytical Chemistry*, 81, 9239–9246, 2009.

Thomas, S. W., III, G. D. Joly, T. M. Swager, Chemical sensors based on amplifying fluorescent conjugated polymers, *Chemical Reviews*, 107, 1339–1386, 2007.

Tsien, R. Y., The green fluorescent protein, *Annual Review of Biochemistry*, 67, 509–544, 1998.

Veinot, J. G. C., Synthesis, surface functionalization, and properties of freestanding silicon nanocrystals, *Chemical Communications*, 40, 4160–4168, 2006.

Wang, X.-D., H.-X. Chen, Y. Zhao, X. Chen, X.-R. Wang, Optical oxygen sensors move towards colorimetric determination, *Trends in Analytical Chemistry*, 29(4), 319–338, 2010a.

Wang, X.-D., R. J. Meier, M. Link, O. S. Wolfbeis, Photographing oxygen distribution, *Angewandte Chemie—International Edition*, 49(29), 4907–4909, 2010b.

Yao, G., L. Wang, Y. R. Wu, J. Smith, J. S. Xu, W. J. Zhao, E. J. Lee, W. H. Tan, FloDots: Luminescent nanoparticles, *Analytical and Bioanalytical Chemistry*, 385, 518–524, 2006.

Yotter, R. A., D. M. Wilson, A review of photodetectors for sensing light-emitting reporters in biological systems, *IEEE Sensors Journal*, 3, 288–303, 2003.

3 Increasing Projector Contrast and Brightness through Light Redirection

Reynald Hoskinson and Boris Stoeber

CONTENTS

3.1 INTRODUCTION

A projection display creates an image on an auxiliary surface, separate from the projector itself. For this reason the images can be made very large, able to be viewed by groups of many people, for instance, at a cinema. Typically projectors are structured with a light source separate from the image-forming element, or *light valve*. As much of the light as possible from the light source is channeled to the light valve, which in turn relays the formed image through the projection optics to the screen.

How efficient the projector is at doing this has major repercussions. Brightness is the primary characteristic determining projector price and quality, and projector efficiency is one key in determining the final brightness of the projected image. Simply increasing the brightness of the lamp to make the image brighter is not always an option. The lamp is typically the most expensive piece of the projector, even more so than the light valve itself. Also, since the light that is not directed to the screen ends up as heat, a brighter lamp carries with it the need for bulkier and noisier fans and lamp electronics.

The light source in a projector supplies a uniform brightness distribution on the light valve. For most images, however, only a fraction of the total area is illuminated at peak brightness. A conventional projector simply blocks the light that is not

49

necessary for a scene, thereby wasting this fraction of light, while it could be used to further illuminate the bright parts of the image. Furthermore, the currently available light valves are "leaky" and cannot block all the light for black image areas (Dewald et al., 2004). The contrast, or dynamic range of a projector, is the ratio between the brightest and darkest levels it can display. Merely increasing the illumination with a more powerful projector lamp does not necessarily increase the dynamic range, as both the darkest and brightest levels rise by the same relative amount. Current projector technology could benefit from an improvement in both dynamic range and peak brightness of the projected image.

The dynamic range of projectors is not sufficient to display many real-world scenes, which can have up to eight orders of magnitude of luminance range (Reinhard et al., 2005). Scenes such as those that include a sunset, fireworks, or daylight must have their luminances tone-mapped to the limited range of current projectors. The peak brightness of a projector also limits the type of environment in which it can be used to its full potential; the brighter the room illumination, the more lower-end detail will be lost, resulting in a "washed out" image.

The emerging popularity of 3D movies even further strains a projector's ability to display an image. Typically the efficiency of a theater projector drops to 14% of what it is for 2D content (Brennesholtz, 2009). Current digital cinema projectors strain to achieve 4.5 foot-lamberts (fl) of screen luminance for a 3D film, especially on larger screens, rather than the 14 fl recommended by the Society of Motion Picture Television Engineers in the standard ANSI/SMPTE 196M (SMPTE, 2003). Even at 4.5 fl, lamps must be run at maximum power, leading to high electric bills and short lamp life. These limitations are felt by the studios as well. Because the displayed luminances are so different, they affect the perception of colors (Brennesholtz, 2009), and so studios must perform two separate expensive color correction processes, one for 2D and one for 3D.

3.2   CONVENTIONAL PROJECTION DISPLAY SYSTEMS

To explain our proposal for a high-contrast projector using an auxiliary mirror array, it helps to first outline the architecture of a modern projector system. Most projection displays available today are based on either digital light projection (DLP) or liquid crystal on silicon (LCoS). They differ in the type of light valve used, the element that selectively blocks light. A lamp provides uniform illumination to the light valve, and the liquid crystals in the LCoS, for example, selectively reduce illumination of a pixel on the screen in order to form the dark parts of the image. The digital micro-mirror device (DMD) inside a DLP projector functions in a similar manner. A DMD is an array of micromirrors, one for each pixel, each one of which can be tilted in one direction so that incident light reflects toward the projection lens and then out onto the screen, or in another direction so the light is reflected to a heat sink and that spot on the screen remains dark.

All projectors can be broken down into a number of functional subsections. Figure 3.1 shows the subsections of single-chip DLP projectors. For purposes of clarity only, the light reaching one pixel of the DMD is shown; in reality, there is a ray bundle that reaches each pixel of the DMD. First in the light path, a reflector collects the light from a small lamp or light-emitting diode (LED) and directs it into

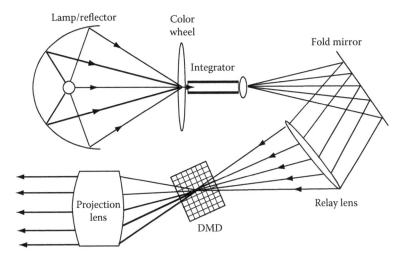

FIGURE 3.1   One ray bundle traversing a projection system.

the illumination optics. In a single-chip DLP projector, the lamp reflector minimizes the spot size of the light at the color wheel. After the color wheel is the integrator, which spatially redistributes the image of the light source from a highly peaked to a more uniform distribution with an aspect ratio that matches that of the light valve. This affects the final distribution of the light on the screen. For DLP projectors, the integrator is usually a rod, made of hollow mirrored tunnels. From the integrator rod, the light travels through relay/folding optics, which form an image of the integrator rod face on the DMD. The image of the DMD is then transmitted to the screen using a projection lens system.

It is interesting to note how the DMD creates the image we see on the screen. Each DMD micromirror has two active tilt positions: plus or minus the maximum tilt angle α_m, with $\alpha_m = 12°$ in current devices. In one state, light is reflected to the projection lens, while a tilt in the other direction sends light to the "light dump," where it is emitted as heat. This places particular constraints on the light incoming to the DMD mirrors. To fulfill its function as a light modulator, the incoming/outgoing light has to be limited to an angle less than $2\,\alpha_m$, $\pm\alpha_m$. That way, the light going to the heat sink can be fully separated from that going to the projection lens pupil. Another way of saying this is that the numerical aperture (NA) of the projection pupil has to be small enough to prevent overlapping flat and on-state light.

3.3   OTHER ADAPTIVE DISPLAY MECHANISMS FOR PROJECTORS

There have been other proposed methods for dynamically changing portions of the display chain in a projector in response to the image being displayed. The addition of a dynamic iris that is adjusted per frame of video is a limited way of adapting the projector's light source to the content (Iisaka et al., 2003). The dynamic iris is a physical aperture near the lamp that partially closes during dark scenes, and opens during bright scenes. While this method can decrease the black level of certain images, it is only a global adjustment, not allowing the improvement of contrast in a scene with several localized bright or dark areas. Also, while this approach decreases the black level for certain select images, it can't increase the maximum brightness of a scene like the analog micromirror array (AMA) can.

There have been several methods proposed that increase contrast solely by decreasing the dark level of the projector, at the expense of brightness. Pavlovych and Stuerzlinger (2005), Damberg et al. (2006), and Kusakabe et al. (2009) each use two light valves in series within a projector system, which reduces the dark level but also the overall brightness and efficiency of the system.

3.4   A CONCEPT FOR AN IMPROVED PROJECTION DISPLAY

To address the contrast shortcomings of currently available projectors, we have proposed adding a low-resolution intermediate mirror array to provide a nonhomogeneous light source (Hoskinson and Stoeber, 2008). This intermediate mirror device is capable of directing the uniform light from the projector lamp incident on its surface to different areas on the light valve, in effect projecting a low-resolution version of the original image onto the light valve, as shown in Figure 3.2. Adding this intermediate mirror device improves the dynamic range in two ways: By directing the light to the bright parts of the image, the achievable peak brightness will be increased. Simultaneously, the amount of light that needs to be blocked in the dark regions of the image is reduced, thus decreasing the brightness of the black level.

This intermediate device can be realized with a low-resolution AMA, made using microelectromechanical system (MEMS) technology. The tip and tilt angle (two degrees of freedom) of the micromirrors in the array can be set continuously in order to direct light to an arbitrary location on the light valve.

3.5   AMA PROJECTOR OPTICAL DESIGN

The AMA must be inserted between the lamp and the DMD in order to change the illumination distribution on the DMD. The problem is then how to achieve the optimal spot size of light from each AMA mirror onto the DMD, while maintaining adequate light coverage of the DMD overall. The image of the DMD with this variable illumination will then be projected to the screen by the projection lens. As well as minimizing the AMA spot size, we would like to maximize the spot displacement for a given mirror tilt angle.

A lens between the AMA and the DMD should optimally relay the light at the proper magnification onto the DMD. To determine the optical lens focal length and placement, we start with the simple lens formula (Hecht, 2002):

$$\frac{1}{f} = \frac{1}{u} + \frac{1}{v}, \tag{3.1}$$

which gives a relationship between distances the focal length of the lens f, the distance between the object and the lens u, and the distance between the lens and the image

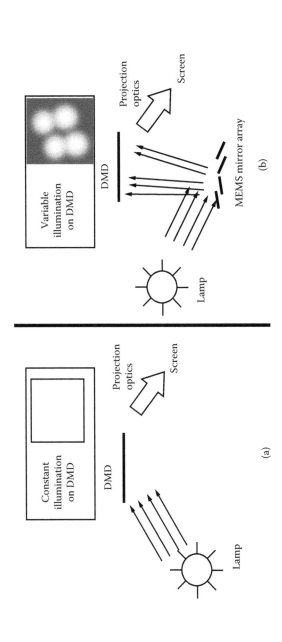

FIGURE 3.2   (a) Schematic of a conventional DLP projector. (b) Schematic of an enhanced DLP projector with second MEMS mirror array (AMA).

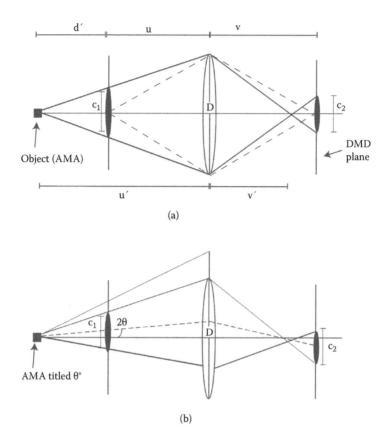

(a)

(b)

FIGURE 3.3 (a) Illustration of circle of confusion. A point on the AMA spreads to a region c_1, which in turn is imaged onto the DMD as the circle c_2. (b) The mirror is tilted by θ, causing the light cone to be shifted by 2θ. The dotted line represents the shifted principal ray of the cone.

v, as shown in Figure 3.3a. If we put the AMA at the object plane u and the DMD at the image plane v, we will get a perfectly in-focus image of the AMA on the screen, minimizing the spot size of the AMA mirror. However, since every point on the object plane is mapped to a corresponding point on the image plane, tilting the mirrors will not move the light from one region to another when the AMA is in focus on the DMD.

To achieve the desired effect of redirecting light from one region to another, the AMA is placed at a distance d' from the object plane of the lens, as shown in Figure 3.3b. By the time the light from a point on the AMA reaches a distance u from the lens, it describes a circle of diameter c_1, which in turn is imaged onto the DMD plane, forming the circle of diameter c_2. This has the effect of blurring the image of the AMA at the DMD plane.

Using plane geometry and knowing the magnification m of the system, we can show that

$$c_2 = \left| Dm\frac{d'}{u'} \right|, \tag{3.2}$$

where D is in effect the lens aperture, limiting the angle of incoming light from the AMA. To lower the rate of increase of blur diameter as the disparity increases, we could reduce the aperture of the lens. However, this would also decrease the efficiency of the projector. To maintain system efficiency, the minimum size of D is that which captures the entire beam from the lamp that can be used by the DMD.

When an AMA mirror is tilted as shown in Figure 3.3b, the incident light is redirected for a distance d' before it reaches the object plane. Points on the plane at u will be imaged onto corresponding points on the plane at v, so the tilt angle of the AMA only displaces light up until d'. The displacement d_t at the DMD plane can thus be calculated as

$$d_t = md'tan(2\theta).$$ (3.3)

From Equations 3.2 and 3.3 it is evident that as we increase the separation d' to increase the displacement of the mobile light (ML) on the DMD, the circle of confusion also grows, so we are blurring the light from the AMA mirror. This relationship is explored in more detail in Hoskinson et al. (2010). A key trade-off of the system is between the blur kernel on one side and the system efficiency and ML range on the other. A smaller blur kernel corresponds to a combination of smaller range and reduced system efficiency.

The distribution of light energy with the circle of confusion can be referred to as the optical point spread function (PSF). Nagahara et al. (2008) provide a model of a PSF that includes an approximation of the effect of aberrations using a Gaussian function,

$$p(r,b) = \frac{2}{\pi(gc)^2} e^{\frac{-2r^2}{(gc)^2}},$$ (3.4)

where g is a constant between 0 and 1. g has a large effect over the shape of the Gaussian, with smaller values giving a more narrow peak. This parameter should be determined through an optical simulation for a given light source, or empirically through optical measurements.

Once we have come up with an estimate of the PSF of one spot on the AMA, we can estimate the distribution of light from one mirror in the array on the DMD, referred to here as the mobile light, by convolving the area of the mirror with the PSF. We can then use the ML to get an estimate of the light distribution from the entire AMA as it appears on the DMD.

Let $g(x, y)$ be the ML, the luminance distribution on the DMD from one AMA micromirror, and $f(x, y)$ be the distribution if the ML was in focus (although still with the displacement due to mirror tilt), and $p(x, y)$ be the point spread function estimated according to Equation 3.4. Then

$$g = p * f,$$ (3.5)

where * is the convolution operator. So to obtain the final light distribution from the AMA, we calculate the displacement and magnification of each AMA mirror, place each mirror's displaced and magnified spot into an image, and then convolve (or multiply, in the Fourier domain) the entire image by the PSF.

3.6 MIRROR ALLOCATION

In an AMA approach, the intensity of light of each mobile light source cannot be changed, but the location on the high-resolution light valve is variable. This means that every ML can be targeted toward the region it is needed. The combination of all the virtual light sources must be sufficient to correctly display the same ratio between intensities as specified in the original image data, but the entire image can be made brighter through reallocation. This section discusses an allocation scheme to take advantage of the flexibility afforded by these MLs to increase the peak brightness of the image. Other schemes are possible (e.g., Hoskinson et al., 2010), but this one is particularly efficient and useful for real-time implementation.

Finding the optimal light distribution for the AMA corresponds to choosing the locations for each of the n MLs from the array of n analog micromirrors. We can approximate the representation of the light incident on the DMD, coming from the entire AMA, as a block of pixels in DMD space. For the sake of example, we will illustrate with an 800×600 pixel DMD. We initially represent the light as a white image of size 1000×800 pixels, centered on the DMD. The extra space around the edges approximates the overfill, which is a common technique in projector design to ensure image uniformity.

The image is subdivided into n rectangles, each of which represents 1 ML. The position of the center of each rectangle in image (DMD) coordinates is recorded. Each rectangle is then blurred using the convolution kernel specified in Equation 3.4, to get the shape and relative intensity distribution of each ML. The convolution kernel used to blur the light is calculated from the physical parameters of the optical system, as detailed above.

Initially, the mirrors of the AMA are in a nonactuated, flat state so all MLs are equally distributed on the DMD. The summation of the MLs in this state should give a relatively uniform distribution on the DMD so that any image can be displayed in a conventional manner. We must then search for the position of each of the n MLs that provide the best improvement to the displayed image. It is assumed that there is adequate granularity in the mirror control to position the MLs at any pixel in the image.

To guide the allocation algorithm, it needs to be established what constitutes an improvement. Here we define improvement as the ability to boost the entire range of pixel brightness values, corresponding to a projector with a higher effective lumen output.

Our approach for allocating the light from all AMA mirrors is to divide the original image into equal energy zones, and allocate one ML for each zone. The median cut algorithm described in Debevec (2005) is an efficient method of subdividing an image into zones of approximately equal energy. We have modified the original algorithm so that it can divide an image into an arbitrary number of regions d. The image is first added to the region list as a single region, along with the number of desired divisions. As long as $d > 2$, subdivide the region into two parts as equally as possible, adding the two new subregions back to the region list, as well as the new divided d for each. In order to maintain the constraint that all regions at the final level be of approximately equal energy, choose the cutting line so that it divides the region into portions of energy as equal as possible, along the largest dimension of the parent region. Repeat until all d's in the region list equal 1.

Figure 3.4 shows an image cut into 28 regions. The centroids of each region are marked with squares. An ML is placed with its center at the centroid of each region. The main advantage of dividing the image in this manner is that the image can be divided quickly into as many regions as there are MLs.

One potential drawback of using this scheme for ML allocation is that the size of each region is a function only of its summed light energy, without regard to the size of the ML. Very small regions might only fit a small fraction of the total light energy of an ML, and very large regions might be larger than a single ML. Also, an equal aspect ratio of a region is not guaranteed, so some regions might be of much different shape than an ML.

Another drawback is that the MLs could have a limited range of movement from their original positions due to a limited tilt angle of the micromirrors, which is not taken into account in the above algorithm. In the Euclidean bipartite minimum matching problem (Agarwal and Varadarajan, 2004), we are given an equal number of points from two sets, and would like to match a point from one set with a distinct point from the other set, so that the sum of distances between the paired points is minimized. We can use an algorithm that solves this problem to match each centroid from the median cut step to a location of an ML in its rest (nontilted) state, and thus minimize the sum total distance between the pairs in the two groups. This will minimize the sum total angle that the mirrors must tilt to achieve the points specified in the median cut solution set. The Euclidean bipartite matching problem can be done in polynomial time using the Hungarian approach (Kuhn, 1955).

FIGURE 3.4 (See color insert.) An image divided into 28 regions using the median cut algorithm, with the centroids in each region represented as dots.

The bipartite minimum matching problem does not limit any one pair to be less than a given amount, only that the sum total of all pairs is minimized. This means that there might be some pairs that exceed the range of motion of an ML for a given maximum mirror tilt angle. If any of the distances between the nontilted ML and the centroid are larger than the range, they are placed at the furthest point along the line that connects the two points that they can reach.

3.7 DESIGN CONSIDERATIONS FOR ELECTROSTATICALLY ACTUATED MICROMIRRORS

In the preceding sections it has become apparent that an analog micromirror array suitable for this application must have specific capabilities. Each micromirror must have adequate range of motion in order to move the light from one region to another. The array must be as efficient as possible so that the losses it introduces do not outweigh the gains. In this section we discuss these requirements and others that must be addressed when designing a micromirror array.

The primary mode of actuation of micromirrors is electrostatic, due to its scalability and low power consumption. Other actuation methods include magnetic (Judy and Muller, 1997) and thermal (Tuantranont et al., 2000) actuation, which are more difficult to confine or require more power, respectively, than electrostatic actuation.

A voltage applied between two separated surfaces creates an attractive electrostatic force (Senturia, 2001). Typically the mobile surface is attached to a spring system that provides a restoring force when the mobile surface (the mirror) approaches the fixed surface.

The electrostatic torque τ_e causes the micromirror to rotate, which in turn causes an opposing mechanical spring torque in the torsion beams suspending the mirror above the substrate. The equilibrium tilt angle is reached when the electrostatic torque equals the opposing mechanical restoring moment:

$$\tau_e = M_t. \tag{3.6}$$

One difficulty of designing micromirrors that are fully controllable over a range of tilt angles is that at one third of the range between the mirror surface and the electrode, the mirror will "pull in," snapping toward the electrode surface. Therefore, care must be taken not to exceed the pull-in value when continuous tilt range is desired. Since gap distance h between the mirror layer and the electrode limits the maximum tilt angle α_1 of a given size mirror, increasing h will extend the tilt angle range, but also increase the voltage required to actuate the mirror.

One other way of increasing α_1 is to decrease the mirror size L_1. Using multiple smaller mirrors ($L_2 > L_1$) in place of large mirrors allows for greater angles ($\alpha_2 > \alpha_1$). However, this also requires many more electrodes to control the direction of each of the small mirrors. It also compromises the fill factor of the array, as the ratio of space between mirrors rises as the mirror size shrinks. Also, more space has to be allocated between mirrors to the electric traces feeding voltages to all of the extra electrodes.

The main example of MEMS micromirrors available in the consumer market is the DMD, which has been discussed above. The DMD chip cannot be used as an

AMA in this application because it has no intermediate positions between its two discrete states. However, both before and after the DMD was invented there has been a substantial amount of research into variable-angle micromirrors for use in optical switching and other applications (Bishop et al., 2002; Tsai et al., 2004b; Dutta et al., 2000; Dokmeci et al., 2004; Taylor et al., 2004).

Applications that require just one mirror, such as laser scanning, offer high tilt angles for single mirrors (Tsang and Parameswaran, 2005). To achieve this, extensive use is made of the chip area around the mirror. This makes such designs unsuitable for applications where multiple closely packed mirrors are required, such as this one.

Some mirror arrays have a high fill factor in one direction only (e.g., Taylor et al., 2004; Tsai et al., 2004a; Hah et al., 2004). The mirrors in these configurations can be stacked tightly in one dimension, but extended components to the sides of the mirrors prevent them from being stacked tightly in two dimensions. Dagel et al. (2006) describe a hexagonal tip/tilt/piston mirror array with an array fill factor of 95%. The micromirrors tilt using a novel leverage mechanism underneath the mirror.

Many tip/tilt mirror systems use gimbals to suspend the mirrors (e.g., Bishop et al., 2002; Lin et al., 2001; Wen et al., 2004). Usually, a frame surrounds the mirror and is attached to it by two torsional springs, forming an axis of rotation. The frame itself is then attached to the surrounding material by two springs in orthogonal directions, allowing the mirror to tip and tilt on torsion springs.

3.7.1 ANALOG MICROMIRROR ARRAY DESIGN

While the above designs meet one or several of the characteristics needed for an AMA, none meet all of them, and in any case, all but the DMD are prototypes, not available to the public. From calculations detailed in Hoskinson et al. (2010), a 7×4 pixel AMA would provide sufficient resolution, and micromirrors with a tilt angle of $\pm 3.5°$ would allow for sufficient light reallocation in any direction. The area of the mirror array should be close to that of the light valve selected, 11.17×8.38 mm for an 800×600 DMD array.

We have designed MEMS micromirrors to use in an AMA projector (Hoskinson et al., 2012) by machining into a 10 μm thick silicon plate that is suspended 13 μm above electrodes on a glass substrate. They employ a thin gimbal system, which suspends square mirrors over the 12 μm cavity. Multiple mirrors are controlled simultaneously to form one AMA composite mirror by linking their electrodes. In our design, mirrors within a row in the same composite mirror share the same gimbal frame, which improves fill factor and promotes homogeneity between submirror deflections. Multiple mirrors within one frame also minimize the ratio between frame area and mirror area. Instead of the loss of reflective area from four frame sides and two frame hinges per mirror, the loss from the frame area and outer hinges is amortized over a larger number of mirrors. The two gimbal springs and two of the four sides are only needed once per group rather than once per mirror.

Figure 3.5 shows a photograph of several of the composite mirrors in the array. Placing a row of mirrors within a gimbal frame also allows the electrodes beneath to be routed in daisy-chain form within the rows, minimizing the electrode paths.

FIGURE 3.5 Array of composite mirrors in Micragem row design.

Thus, there is a series of rectangular rows of mirrors within a composite mirror, with corresponding rows in the electrode layer.

3.8 OPTICAL SYSTEM

A prototype to demonstrate dual light modulation using an AMA and DMD has been built using two conventional projectors and a custom-made AMA chip. The light from one projector lamp is collected with a 60 mm lens and directed to the AMA. A 45 mm lens relays the light from the AMA to a second projector, a Mitsubishi PK20 pico-projector that is relatively easy to open up to allow access to the DMD. The light incident to the DMD is reflected normally through a prism to the PK20 projection lens onto a screen. Figure 3.6 shows a photograph of the prototype. Physically, the AMA array is 9×5 mm, while the DMD is 11.17×8.38 mm.

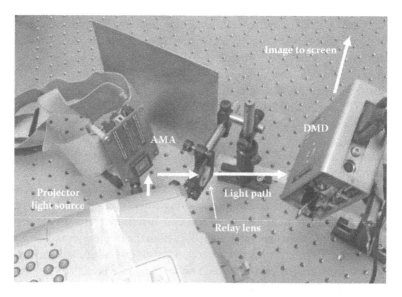

FIGURE 3.6 Photograph of prototype, including projector light source, AMA, relay lens, and DMD.

A custom-made driver provides 84 independent analog voltages to the AMA. A static image was displayed with the AMA projector by directing a video graphics array (VGA) signal from a computer to both the light source and the DMD. Actuating the AMA mirrors via the 84-channel d/a converter changes the final intensity distribution of the projected image.

In the prototype, the light reflecting from the AMA to the DMD and finally out of the projection lens was the sole source of illumination for the images. The disparity can be adjusted by moving the relay lens between AMA and DMD. As the blur increases, details of the AMA disappear, and the resulting ML becomes more and more disperse even as the displacement of the ML increases, as shown in Figure 3.7. The grid lines are the image sent to the DMD, and are in 50-pixel increments. They serve as a reference to estimate the size of the AMA MLs that get larger as the disparity increases because of the increasing size of the blur kernel.

Figure 3.7a shows the ML nearly in focus, with a slight bias in light to the top right. At such low disparities, the ML hardly moves at all. The size of the ML is approximately 75 DMD pixels per side. If the DMD had been exactly on the focal plane, the ML would not have been displaced at all. Figure 3.7b shows how the ML has become blurred as the disparity increased to 17 mm. The size of the ML has grown to 100 pixels per side. The ML has been displaced 60 pixels in a diagonal direction toward the bottom left. The angle of tilt is not uniform over the composite mirror, causing the ML to distort as it is tilted.

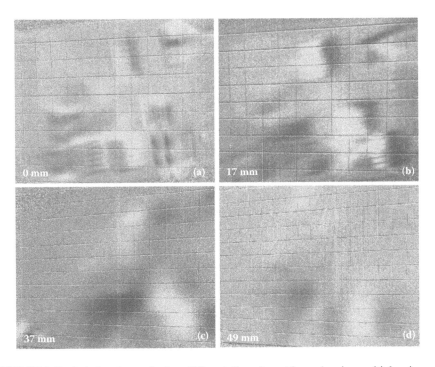

FIGURE 3.7 Relative change for four different disparity settings, showing multiple mirrors actuated.

At 37 mm disparity, the size of the ML is now 120 pixels on its longest side, as shown in Figure 3.7c. The ML is spread over an area of 160 pixels per side. In this case, the ML has become so diffuse that the benefit from the AMA is much less apparent. That is even more the case with Figure 3.7d, which shows roughly the same light distribution at 49 mm.

Overall, these measurements show that the approach of using an AMA is successful at redistributing light from one region of the projected image to another. We also demonstrated that the AMA does not geometrically distort the image from the DMD in any way. The effect that disparity has on both range and blur was shown, and it was demonstrated that the AMA can make areas of the image both brighter and darker as the mirrors are actuated.

3.9   CONCLUSION

By channeling light to where it is needed and away from where it is not, an AMA projector makes better use of its light source, which is one of the most expensive components of today's projectors. After the addition of an AMA, the light reaching the primary image modulator, such as a DMD, is not considered to be uniform. Instead, the distribution from the AMA mirrors is simulated, and the compensation for the nonhomogeneity is applied to the original image before it is sent to the DMD. The result is an image of higher contrast and peak brightness than would otherwise be possible with the same projection lamp.

This work has demonstrated that an AMA-enhanced projector can make projectors intelligent allocators of their light sources. This ability opens up several interesting research directions, as the projector could conceivably take into account the ambient lighting in the room, the psychophysical limitations of the viewers, and the content being presented to display the best possible image over a wide range of conditions.

REFERENCES

Agarwal, P., and K. Varadarajan. A near-linear constant-factor approximation for Euclidean bipartite matching? In *Proceedings of the Twentieth Annual Symposium on Computational Geometry*, New York, NY, 2004, 247–252.

Bishop, D.J., C. Randy Giles, and G.P. Austin. The Lucent LambdaRouter: MEMS technology of the future here today. *IEEE Communications Magazine* 40, 3: (2002) 75–79.

Brennesholtz, M.S. 3D: Brighter is better. 2009. http://www.widescreenreview.com/news_detail.php?id=18500.

Dagel, D.J., W.D. Cowan, O.B. Spahn, G.D. Grossetete, A.J. Grine, M.J. Shaw, P.J. Resnick, and B. Jokiel Jr. Large-stroke MEMS deformable mirrors for adaptive optics. *Journal of Microelectromechanical Systems* 15, 3: (2006) 572–583.

Damberg, G., H. Seetzen, G. Ward, W. Heidrich, and L. Whitehead. High dynamic range projection systems. Presented at Proceedings of the 2006 Society for Information Display Annual Symposium, San Francisco, 2006.

Debevec, P. A median cut algorithm for light probe sampling. Presented at SIGGRAPH '05: ACM SIGGRAPH 2005 Posters, New York, 2005.

Dewald, D.S., D.J. Segler, and S.M. Penn. Advances in contrast enhancement for DLP projection displays. *Journal of the Society for Information Display* 11: (2004) 177–181.

Dokmeci, M.R., A. Pareek, S. Bakshi, M. Waelti, C.D. Fung, K. Hang Heng, and C.H. Mastrangelo. Two-axis single-crystal silicon micromirror arrays. *Journal of Microelectromechanical Systems* 13, 6: (2004) 1006–1017.

Dutta, S.B., A.J. Ewin, M.D. Jhabvala, C.A. Kotecki, J.L. Kuhn, and D.B. Mott. Development of individually addressable micromirror arrays for space applications. *Proceedings of SPIE* 4178: (2000) 365–371.

Hah, D., S. Ting-Yu Huang, J.-C. Tsai, H. Toshiyoshi, and M.C. Wu. Low-voltage, large-scan angle MEMS analog micromirror arrays with hidden vertical comb-drive actuators. *Journal of Microelectromechanical Systems* 13, 2: (2004) 279–289.

Hecht, E. *Optics.* 4th ed. San Francisco: Addison Wesley, 2002.

Hoskinson, R., S. Hampl, and B. Stoeber. Arrays of large-area, tip/tilt micromirrors for use in a high-contrast projector. *Sensors and Actuators A: Physical* 173, 1: (2012) 172–179.

Hoskinson, R., and B. Stoeber. High-dynamic range image projection using an auxiliary MEMS mirror array. *Optics Express* 16: (2008) 7361–7368.

Hoskinson, R., B. Stoeber, W. Heidrich, and S. Fels. Light reallocation for high contrast projection using an analog micromirror array. *ACM Transactions on Graphics (TOG)* 29, 6: (2010) 165.

Iisaka, H., T. Toyooka, S. Yoshida, and M. Nagata. Novel projection system based on an adaptive dynamic range control concept. In *Proceedings of the International Display Workshops*, 2003, vol. 10, 1553–1556.

Judy, J.W., and R.S. Muller. Magnetically actuated, addressable microstructures. *Journal of Microelectromechanical Systems* 6, 3: (1997) 249–256.

Kuhn, H.W. The Hungarian method for the assignment and transportation problems. *Naval Research Logistics Quarterly* 2: (1955) 83–97.

Kusakabe, Y., M. Kanazawa, Y. Nojiri, M. Furuya, and M. Yoshimura. A high-dynamic-range and high-resolution projector with dual modulation. *Proceedings of SPIE* 7241: (2009).

Lin, J.E., F.S.J. Michael, and A.G. Kirk. Investigation of improved designs for rotational micromirrors using multi-user MEMS processes. *SPIE MEMS Design, Fabrication, Characterization, and Packaging* 4407: (2001) 202–213.

Nagahara, H., S. Kuthirummal, C. Zhou, and S.K. Nayar. Flexible depth of field photography. Presented at European Conference on Computer Vision (ECCV), 2008.

Pavlovych, A., and W. Stuerzlinger. A high-dynamic range projection system. *Progress in Biomedical Optics and Imaging* 6: (2005) 39.

Reinhard, E., G. Ward, S. Pattanaik, and P. Debevec. *High dynamic range imaging: Acquisition, display, and image-based lighting.* Morgan Kaufmann Series in Computer Graphics. San Francisco: Morgan Kaufmann, 2005.

Senturia, S.D. *Microsystem design.* Norwell, MA: Kluwer Academic, 2001.

SMPTE. 196M indoor theater and review room projection—Screen luminance and viewing conditions. 2003.

Taylor, W.P., J.D. Brazzle, A. Bowman Osenar, C.J. Corcoran, I.H. Jafri, D. Keating, G. Kirkos, M. Lockwood, A. Pareek, and J.J. Bernstein. A high fill factor linear mirror array for a wavelength selective switch. *Journal of Micromechanics and Microengineering* 14, 1: (2004) 147–152.

Tsai, J.-C., L. Fan, D. Hah, and M.C. Wu. A high fill-factor, large scan-angle, two-axis analog micromirror array driven by leverage mechanism. Presented at International Conference on Optical MEMS and Their Applications, Japan, 2004a.

Tsai, J.-C., S. Huang, and M.C Wu. High fill-factor two-axis analog micromirror array for $1xN^2$ wavelength-selective switch. In *IEEE International Conference on Micro Electro Mechanical Systems*, 2004b, 101–104.

Tsang, S.-H., and M. Parameswaran. Self-locking vertical operation single crystal silicon micromirrors using silicon-on-insulator technology. In *Canadian Conference on Electrical and Computer Engineering 2005*, 2005, 429– 432.

Tuantranont, A., V.M. Bright, L.A. Liew, W. Zhang, and Y.C. Lee. Smart phase-only micromirror array fabricated by standard CMOS process. In *Thirteenth Annual International Conference on Micro Electro Mechanical Systems 2000 (MEMS 2000)*, 2000, 455–460.

Wen, J., X.D. Hoa, A.G. Kirk, and D.A. Lowther. Analysis of the performance of a MEMS micromirror. *IEEE Transactions on Magnetics* 40: (2004) 1410–1413.

4 Vision-Aided Automated Vibrometry for Remote Audio, Visual, and Range Sensing

Tao Wang and Zhigang Zhu

CONTENTS

4.1 INTRODUCTION

Remote object signature detection is becoming increasingly important in noncooperative and hostile environments for many applications (Dedeoglu et al., 2008; Li et al., 2008). These include (1) remote and large area surveillance in frontier defense, maritime affairs, law enforcement, and so on; (2) perimeter protection for important locations and facilities such as forests, oil fields, railways, and high-voltage towers; and (3) search and rescue in natural and man-made disasters such as earthquakes, flooding, hurricanes, and terrorism attacks. In these situations, target signature detection, particularly signatures of humans, vehicles, and other targets or events, at a large distance is critical in order to watch out for the trespassers or events before taking appropriate actions, or make quick decisions to rescue the victims, with minimum risks. Although imaging and video technologies (including visible and infrared (IR)) have had great advancement in object signature detection at a large distance, there are still many limitations in noncooperative and hostile environments because of intentional camouflage and natural occlusions. Audio information, another important data source for target detection, still cannot match the range and signal qualities provided by video technologies for long-range sensing, particularly under a variety of large background noises. For obtaining better performance of human tracking in a near to mediate range, Beal et al. (2003) and Zou and Bhanu (2005) have reported the integrations of visual and acoustic sensors. By integration, each modality may compensate for the weaknesses of the other one. But in these systems, the acoustic sensors (microphones) need to be placed near the subjects in monitoring, and therefore cannot be used for long-range surveillance. A parabolic microphone, which can capture voice signals at a fairly large distance in the direction pointed by the microphone, could be used for remote hearing and surveillance. But it is very sensitive to noise caused by the surroundings (i.e., wind) or the sensor motion, and all the signals on the way are captured. Therefore, there is a great necessity to find a new type of acoustic sensor for long-range voice detection.

Laser Doppler vibrometers (LDVs) such as those manufactured by Polytec (2010) and Ometron (2010) can effectively detect vibration within 200 m with sensitivity in the order of 1 μm/s. Larger distances could be achieved with the improvements of sensor technologies and the increase of the laser power while using a different wavelength (e.g., infrared instead of visible). In our previous work (Li et al., 2006; Zhu et al., 2007), we have presented very promising results in detecting and enhancing voice signals of people from large distances using a Polytec LDV. However, the user had to manually adjust the LDV sensor head in order to aim the laser beam at a surface that well reflects the laser beam, which was a tedious and difficult task. In addition, it was very hard for the user to see the laser spot at a distance above 20 m, and so it was extremely difficult for the human operator to aim the laser beam of the LDV on a target at a distance larger than 100 m. Of course, human eyes cannot see infrared laser beams, so it would be a serious problem if the LDV used infrared. Also, it takes quite some time to focus the laser beam, even if the laser beam is pointed to the surface. Therefore, reflection surface selection and automatic laser aiming and focusing are greatly needed in order to improve the performance and efficiency of the LDV for long-range hearing.

Here, we present a novel multimodal sensing system, which integrates the LDV with a pair of pan-tilt-zoom (PTZ) cameras to aid the LDV in finding a reflective surface and focusing its laser beam automatically, and consequently, the system captures both video and audio signals synchronously for target detection using multimodal information. In addition to video and audio, this sensing system can also obtain range information using LDV-PTZ triangulation as well as stereo vision using the two cameras. The range information will further add values to object signature detection in addition to the audio and video information, and improve the robustness and detection rate of the sensor. The main contribution of this work is the collaborative operation of a dual-PTZ-camera system and a laser pointing system for long-range acoustic detection. To our knowledge, this is the first work that uses a PTZ stereo for automating the long-range laser-based voice detection. Meanwhile, the combination is a natural extension of the already widely used PTZ-camera-based video surveillance system toward multimodal surveillance with audio, video, and range information.

The rest of the chapter is organized as follows. Section 4.2 presents some background and related work. Section 4.3 describes an overview of our vision-aided automated vibrometry system. Section 4.4 discusses the calibration issues among the multimodal sensory components. Section 4.5 shows the algorithms for feature matching and distance measuring using the system. Section 4.6 describes the adaptive and collaborative sensing approach. Section 4.7 provides some experimental results. Finally, we conclude our work in Section 4.8.

4.2 BACKGROUND AND RELATED WORK

4.2.1 PRINCIPLE OF THE LASER DOPPLER VIBROMETER

The LDV works according to the principle of laser interferometry. Measurement is made at the point where the laser beam strikes the structure under vibration. In the heterodyning interferometer (Figure 4.1), a coherent laser beam is divided into object and reference beams by a beam splitter BS1. The object beam strikes a point on the moving (vibrating) object, and light reflected from that point travels back to beam splitter BS2 and mixes (interferes) with the reference beam at beam splitter BS3. If the object is moving (vibrating), this mixing process produces an intensity fluctuation in the light as

$$I_1 = \frac{1}{2}A^2\left\{1-\cos\left[2\pi\left(f_B+\frac{2v}{\lambda}\right)t\right]\right\} \tag{4.1}$$

where I_1 is light intensity, A is the amplitude of the emitted wave, f_B is modulation frequency of the reference beams, λ is the wavelength of the emitted wave, v is the object's velocity, and t is observation time. A detector converts this signal

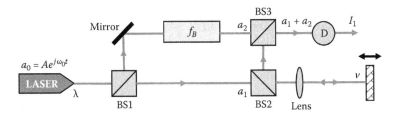

FIGURE 4.1 Principle of the laser Doppler vibrometer.

to a voltage fluctuation. And from the fluctuating of light patterns, the velocity of the object can be decoded by a digital quadrate demodulation method (Scruby and Drain, 1990). An interesting finding of our study is that most objects vibrate while wave energy (including that of voice waves) is applied to them. Although the vibration caused by the voice energy is very small compared with other vibration, it can be detected by the LDV, and be extracted with advanced signal filtering. The relation of voice frequency f, velocity v, and magnitude m of the vibration is

$$v = 2\pi f m \qquad (4.2)$$

As seen from the above principle of the LDV, there are three requirements to be considered in order to use the LDV to measure the vibration of a target caused by sounds:

1. An appropriate surface close to the sounding target with detectable vibration and good reflection index.
2. The focus of the LDV laser beam on the refection surface; otherwise, very weak reflection signals are obtained due to the scattering of coherent light and path length differences.
3. A necessary signal enhancement process to filter out the background noise and the inherent noise of the LDV.

In close-range and lab environments, it is not a serious problem for a human operator to find an appropriate reflective surface, focus the laser beam, and acquire the vibration signals. But at a large distance (from 20 to 100 m), the manual process becomes extremely difficult because it is very hard for a human operator to aim the laser beam to a good reflective surface. Also, it takes quite some time to focus the laser beam, even if the laser beam is pointed to the surface. Therefore, there are great unmet needs in facilitating the process of surface detection, laser aiming, laser focusing, and signal acquisition of the emerging LDV sensor, preferably through system automation.

4.2.2 RELATED WORK

Acoustic sensing and event detection can be used for audio-based surveillance, including intrusion detection (Zieger et al., 2009) and abnormal situation detection in public areas, such as banks, subways, airports, and elevators (Clavel et al., 2005; Radhakrishnan et al., 2005). It can also be used as a complementary source of information for video surveillance and tracking (Cristani et al., 2007; Dedeoglu et al., 2008). In addition to microphones, an LDV, as another type of acoustic sensor, is a novel type of measurement device to detect a target's vibration in a noncontact way, in applications such as bridge inspection (Khan et al., 1999), biometrics (Lai et al., 2008), and underwater communication (Blackmon and Antonelli, 2006). It has also been used to obtain the acoustic signals of a target (e.g., a human or a vehicle) at a large distance by detecting the vibration of a reflecting surface caused by the sound of the target next to it (Zhu et al., 2005, 2007; Li et al., 2006; Wang et al., 2011a). The LDVs have been used in the inspection industry and other important applications concerning environment, safety, and preparedness that meet basic human needs. In bridge and building inspection, the noncontact vibration measurements for monitoring structural defects eliminate the need to install sensors as a part of the infrastructure (e.g., Khan et al., 1999). In security and perimeter applications, an LDV can be used for voice detection without having the intruders in the line of sight (Zhu et al., 2005). In medical applications, an LDV can be used for noncontact pulse and respiration measurements (Lai et al., 2008). In search and rescue scenarios where reaching humans can be very dangerous, an LDV can be applied to detect survivors that are out of visual sight. Blackmon and Antonelli (2006) have tested and shown a sensing system to detect and receive underwater communication signals by probing the water surface from the air, using an LDV and a surface normal tracking device.

However, in most of the current applications, such systems are manually operated. In close-range and lab environments this is not a very serious problem. But in field applications, such as bridge/building inspection, area protection, or search and rescue applications, the manual process takes a very long time to find an appropriate reflective surface, focus the laser beam, and get a vibration signal—more so if the surface is at a distance of 100 m or more. A vision-aided LDV system can improve the performance and efficiency of the LDV for automatic remote hearing. In this work, we improved the flexibility and usability of the vision-aided automated vibrometry system from our previous design with a single PTZ camera (Qu et al., 2010) to the current design with a pair of PTZ cameras (in Section 4.3) and by providing adaptive and collaborative sensing (in Section 4.6).

4.3   VISION-AIDED AUTOMATED VIBROMETRY: SYSTEM OVERVIEW

The system consists of a single-point LDV sensor system, a mirror mounted on a pan-tilt unit (PTU), and a pair of PTZ cameras, one of which is mounted on the top of the PTU (Figure 4.2). The sensor head of the LDV uses a helium-neon laser with

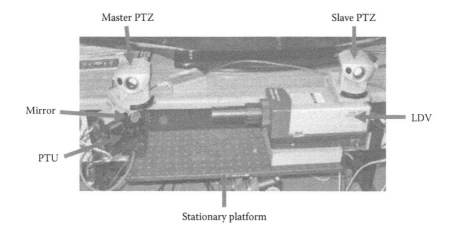

FIGURE 4.2   The multimodal sensory platform.

a wavelength of 632.8 nm and is equipped with a super long-range lens. It converts velocity of the target into interferometry signals and magnitude signals, and sends them to the controller of the LDV, which is controlled by a computer via an RS-232 port. The controller processes signals received from the sensor head of the LDV, and then outputs either voltage or magnitude signals to the computer using an S/P-DIF output. The Polytec LDV sensor OFV-505 and the controller OFV-5000 that we use in our experiments can be configured to detect vibrations under several different velocity ranges: 1, 2, 10, and 50 mm/s/V, where V stands for velocity. For voice vibration of a basic frequency range of 300 to 3000 Hz, we usually use the 1 mm/s/V velocity range. The best resolution is 0.02 μm/s under the range of 1 mm/s/V according to the manufacturer's specification with retroreflective tape treatment. Without the retroreflective treatment, the LDV still has a sensitivity on the order of 1.0 μm/s. This indicates that the LDV can detect vibration (due to voice waves) at a magnitude in nanometers without retroreflective treatment; this can even get down to picometers with retroreflective treatment.

The LDV sensor head weighs about 3.4 kg; this is the major reason that a mirror mounted on the PTU is used in our system to reflect the laser beam to freely and quickly point it to various directions in a large field of view. The laser beam points to the mirror at the center of the panning-tilting of the PTU. The vision component consists of a pair of Canon VC-C50i (26×) PTZ cameras, with one mounted on the top of the PTU, which is called the *master PTZ* since it is the main camera to track the laser beam, and the other one mounted on the top of the LDV, which is called the *slave PTZ*. Each PTZ camera (Canon VC-C50i) has a 720×480 focal plane array and an auto-iris zoom lens that can change from 3.5 to 91 mm (26× optical power zoom). The pan angle of the PTZ is ±100° with a rotation speed of 1–90°/s, and the tilt angle of it is from −30° to +90° with a rotation speed of 1–70°/s. The PTU is the model PTU-D46-70 of Directed Perception, Inc. It has a pan range from −159° to +159° and a tilt range from −47° to +31°. Its rotation resolution is 0.013° and its max rotation speed is 300°/s. The reason to use zoom cameras is to detect targets and assist the laser pointing and focusing at various distances. However, at a long distance, the laser spot is usually hard to see with the cameras, either zoomed or with wide views, if the laser is unfocused or not pointed on the right surface. Therefore, the master PTZ camera is used to rotate synchronously with the reflected laser beam from the mirror in order to track the laser spot. Although the laser point may not be observed from the master PTZ, we always control the pan and tilt angles of the master PTZ camera so that its optical axis is parallel to the reflected laser beam, and therefore the laser spot is always close to the center of the image. Then, the master PTZ camera and the slave PTZ form a stereo vision system to obtain the distance to focus the laser spot as well as guide the laser to the right surface for acoustic signal collection. The baseline between the two PTZ cameras is about 0.6 m for enabling long-range distance measurements. In order to obtain the distance from the target surface to the LDV, the calibration among the two PTZ cameras and the LDV is the first important step, which will be elaborated upon in the next section before the discussion of our method for distance measurement.

4.4   SYSTEM CALIBRATION: FINDING PARAMETERS AMONG THE SENSOR COMPONENTS

There are two stereo vision components in our system: stereo vision between the two PTZ cameras and stereo triangulation between the slave PTZ camera and the mirrored LDV laser projection. The first component is used to obtain the range of a point in a reflective surface by matching its image projection (x, y) in the master camera to the corresponding image point (x', y') in the slave camera. The second component is mainly used to obtain the pan (α) and tilt (β) rotation angles of the PTU so that the LDV points to the image point (x, y) in the master image. Before determining the distance, several coordinate systems corresponding to the multisensory platform (Figure 4.2) are illustrated in Figure 4.3 (left): the master PTZ camera coordinate system (S_c), the slave PTZ camera coordinate system $(S_{c'})$, the LDV coordinate system (S_L), and the PTU coordinate system (S_u).

We assume the mirror coordinate system is the same as the PTU coordinate system since the laser will point to the mirror at the origin of the PTU system. The mirror normal direction is along the Z_u axis and initially points to the outgoing laser beam along Z_l. In order to always actually track the reflected laser beam, visible or invisible (by having the optical axis of the master PTZ parallel to the reflected laser beam), the master PTZ not only rotates the same base angles with the PTU, α and β, which are the pan and tilt angles of the PTU around the X_u and Z_u axes, but also undergoes additional pan and tilt rotations $(\alpha'$ and $\beta')$ around the Y_c and X_c axes. We will explain in detail how to determine these angles in Section 4.6.

The stereo matching is performed after the full calibration of the stereo component of the two PTZ cameras, and that between the slave PTZ camera and the "mirrored" LDV. Given a selected point on a reflective surface in the image of the master camera, we first find its corresponding point in the image of the slave camera. Meanwhile, we calculate the pan and tilt angles of the PTU and the master and slave PTZ cameras so that the laser spot is right under the center of the image of the master PTZ camera; the offset to the center is a function of the distance of the surface to the sensor system. The farther the surface is, the closer is the laser spot to the center. The distance from the target point to the optical center of the LDV is estimated via the stereo PTZ and then used to focus the laser beam to the surface.

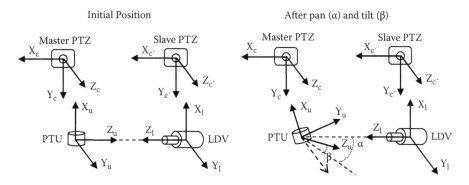

FIGURE 4.3   Coordinate systems of the multimodal platform.

4.4.1 CALIBRATION OF THE TWO PTZ CAMERAS

The calibration between the two PTZ cameras is carried out by estimating both the intrinsic and extrinsic parameters of each camera on every possible zoom factor when the camera is in focus, using the same world reference system. We use the calibration toolbox by Bouguet (2008) to find a camera's parameters under different zoom factors. We have found that the estimated extrinsic parameters do not change much with the changes of zooms. However, the focal lengths of the cameras increase nonlinearly with the changes of different zooms; therefore, we have calibrated the camera under every possible zoom. Also note that the focal lengths of two PTZ cameras may not be the same under the same zoom factor. In order to achieve similar fields of view (FOVs) and to ease the stereo matching between two images, the correct zoom of the slave PTZ camera corresponding to the actual focal length of the master PTZ camera should be selected. After the calibration, we obtain the effective focal lengths and image centers of the two cameras under every zoom factor k, and the transformation between the two cameras, represented by R and T:

$$P_{C'} = RP_c + T \tag{4.3}$$

where P_c and $P_{c'}$ are the representations of a 3D point in the master and slave PTZ coordinate systems (S_C and $S_{C'}$), respectively.

4.4.2 CALIBRATION OF THE SLAVE CAMERA AND THE LDV

Since the intrinsic parameters of the slave PTZ camera have been obtained previously, we only need to estimate the extrinsic parameters characterizing the relation between the LDV coordinate system (S_L) and the slave PTZ camera coordinate system ($S_{C'}$), defined as

$$P_L = R_{C'}P_{C'} + T_{C'} \tag{4.4}$$

where P_L and $P_{C'}$ represent the coordinates of a 3D point in S_L and $S_{C'}$, respectively. $R_{C'}$ and $T_{C'}$ are the rotation matrix and translation vector between S_L and $S_{C'}$. Next, the relation between the LDV coordinate system (S_L) and the mirrored LDV coordinate system (S_{ML}, not shown in Figure 4.3) is defined as

$$P_L = R_U R_{LR} R_U^T (P_{ML} - T_U) + T_U \tag{4.5}$$

where P_L and P_{ML} are the 3D point representations in the S_L and S_{ML}, respectively, and R_U and T_U are the rotation matrix and translation vector between S_L and the PTU coordinate system. The R_{LR} is the rotation matrix that converts a right-hand coordinate system to a left-hand coordinate system. Then the extrinsic parameters are estimated by combining Equations 4.4 and 4.5:

$$R_{C'}P_{C'} = R_U R_{LR} R_U^T (P_{ML} - T_U) + (T_U - T_{C'}) \tag{4.6}$$

For the calibration between the LDV and the slave PTZ, the LDV laser beam is projected at preselected points in a checkerboard placed at various locations/orientations.

Because both the variables $P_{ML} - T_U$ and $T_U - T_{C'}$ are not independent in Equation 4.6, the distance between the fore lens of the LDV and the laser point on the mirror is estimated initially. Also, to avoid the complexity of the nonlinear equation, we assume the initial rotation matrix is an identity matrix, which can be manually adjusted by pointing both cameras parallel to the same direction. Then this initial distance and initial rotation matrix can be refined iteratively. Given n 3D points, $3n$ linear equations that include $n + 14$ unknowns are constructed using Equation 4.6. Therefore, at least seven points are needed.

4.5 STEREO VISION: FEATURE MATCHING AND DISTANCE MEASURING

4.5.1 Stereo Matching

After calibration, the distance of a point can be estimated when the corresponding point in the slave image of a selected point in the master image is obtained. In the master camera, a target point can be selected either manually or automatically. We assume that both left and right images can be rectified given the intrinsic matrices for both cameras and the rotation matrix and translation vector. So, given any point $(x, y, 1)$T in the original (right) image, the new pixel location $(x', y', 1)$T in the rectified right image is $R'_r (x, y, 1)$T. To simplify the task, radial distortion parameters are ignored. The rectified matrices for both cameras (virtually) make both camera image planes the same. Thus, the stereo-matching problem turns into a simple horizontal searching problem since all epipolar lines are parallel. For example, in Figure 4.4, the right image is captured by the master PTZ camera and the left image by the slave PTZ camera. The same target points are shown in white circles. The epipolar line is shown in green across both images. Note that due to the calibration error, the corresponding point may not be exactly on the epipolar line. To solve this problem, a small search window is used to match the region around the selected point with a small range in the vertical direction as well. Since we are only interested in the selected point on a particular reflective surface, this step is very fast.

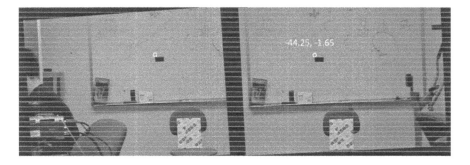

FIGURE 4.4 Stereo matching of the corresponding target points on the images of the master camera (right) and slave camera (left). The numbers on the right image show the pan and tilt angles of the PTU in order to point the laser beam to the target point.

4.5.2 DISTANCE MEASURING

Once two corresponding points lying on the same horizontal epipolar line are identified, the distance can be calculated based on triangulation using the baseline (B) of two rectified cameras. The relation between B and the range (D) of the target surface represented in the master camera system is defined as

$$\frac{B}{D} = \left[x_r \ -x_l \right] \left[\frac{1}{F_{xr}} \ \frac{1}{F_{xl}} \right]^{T} \tag{4.7}$$

where x_r and x_l are the x coordinates of the selected point in the right and left image, and F_{xr} and F_{xl} are the focal lengths of the two PTZ cameras. Ideally, both PTZ cameras should have the same focal length after adjusting their zoom factors.

The calibration result of the slave camera and the LDV is mainly used to determine the pan (α) and tilt (β) angles of the PTU in order to direct the laser beam to the selected point. The conventional triangulation method [16] is used to match the ray from the optical center of the PTZ to the ray of the reflected laser beam. Then the corresponding pan and tilt angels are estimated. Figure 4.4 shows an example of the calculated pan and tilt angels (on the right image) corresponding to the point (white circle) in the left image. Then, given the pan and tilt rotations of the PTU and knowing the corresponding 3D point in the slave camera system as

$$P_{C'} = R'[P_{C'X}, P_{C'Y}, D]^{T} \tag{4.8}$$

where R' is the pan and tilt rotation of the slave PTZ, initially it equals an identity matrix if the slave PTZ camera is in its initial pose when it was calibrated. The estimated LDV distance $D_L = \|P_{ML}\|$ can be then defined based on Equation 4.6, which will be used for focusing the laser beam to the target point.

4.6 ADAPTIVE AND COLLABORATIVE SENSING

The overall goal of this system is to acquire meaningful audio signatures with the assistance of video cameras by pointing and focusing the laser beam to a good surface. However, a target location either manually or automatically selected may not return signals with a sufficient signal-to-noise ratio (SNR). Then a reselection of new target points is required. Figure 4.5 shows the basic idea of adaptive sensing, or adaptively adjusting the laser beam based on the feedback of its returned signal levels.

The stereo matching here is used for obtaining the target distance to the system platform, and then we can automatically focus the laser point to the selected target. This involves the following procedures:

1. A point on a surface close to a designated target is selected either manually or automatically.
2. The target range and the distance from the point to the optical center of the LDV are measured.

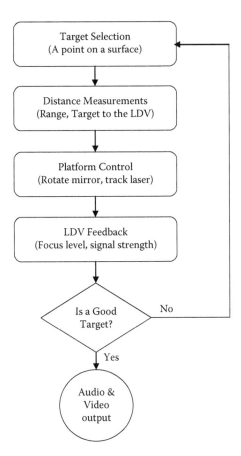

FIGURE 4.5 Flowchart of adaptive sensing for laser pointing and tracking for audio and video signature acquisition.

3. The laser spot is moved to the new location, and the master PTZ camera is rotated synchronously to put the laser spot in the center of images.
4. The laser beam of the LDV is automatically and rapidly focused based on the estimated distance and signal levels, as we did in Qu et al. (2010).

If the selected target point does not have sufficient good returning signals for voice detection, we need to reselect new target points. If the target point is good enough, we can use it to record the audio signature as well as video signatures. In this procedure, there are two key issues that need to be emphasized. First, what is a good surface and how do we select a surface? Second, how do we align the laser beam with the optical center of the camera accurately?

4.6.1 SURFACE SELECTION

The selection of reflection surfaces for LDV signals is important since it is a major factor that determines the quality of acquired vibration signals. There are two basic requirements for a good surface: vibration to the voice energy and reflectivity to the helium-neon laser. We have found that almost all natural objects vibrate more or less with normal sound waves. Therefore, the key technique in finding a good reflection surface is to measure its reflectivity. Based on the principle of the LDV sensor, the relatively poor performance of the LDV on a rough surface at a large distance is mainly due to the fact that only a small fraction of the scattered light (approximately one speckle) can be used because of the coherence consideration. A stationary, highly reflective surface usually reflects the laser beam of the LDV very well. Unfortunately, the body of a human subject does not have such good reflectivity to obtain LDV signals unless (1) it is treated with retroreflective materials and (2) it can keep still relative to the LDV. Also, it is hard to have a robust signal acquisition on a moving object. Therefore, background objects near the interested target are selected and compared in order to detect useful acoustic signals. Typically, a large and smooth background region that has a color close to red is selected for the LDV pointing location.

4.6.2 LASER-CAMERA ALIGNMENT

The next issue is how to automatically aim and track the laser spot, especially for long-range detection. The laser spot may not be observable at a long range, particularly if it is not focused or it does not point on the surface accurately. We solve this problem by always keeping the reflected laser beam parallel to the optical axis of the master PTZ camera so that the laser spot is right under and very close to the center of the master image. Figure 4.6 shows a typical example of a laser spot that is right below the image center in few pixels. We make the ray from the optical center of the master PTZ camera parallel to the reflected laser beam by rotating the PTZ camera with the PTU synchronously (since the PTZ is mounted on the PTU), and then with additional PTZ camera rotations.

The main issue now is how to obtain the additional pan (α') and tilt (β') angles of the PTZ camera given the pan (α) and tilt (β) angles of the PTU. Figure 4.7 shows the relationship between the outgoing laser beam from the LDV ($\overrightarrow{BA}$) and the reflected laser ray ($\overrightarrow{AD}$), with the mirror normal ($\overrightarrow{AC}$). By projecting the reflected ray and

FIGURE 4.6 Two examples of laser point tracking. Both images show the same cropped size (240×160) around the image center with a focused laser spot close to it (black-white dot). The laser point on a white board (indoor, about 6 m) in the left image is closer to the LDV; therefore, it is closer to the image's center than the laser point on a metal pole (outdoor, about 100 m) in the right image.

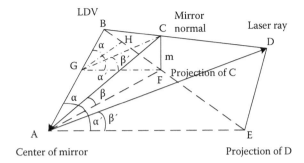

FIGURE 4.7   Geometry model of laser beam from the LDV ($\overrightarrow{BA}$) and its reflected laser ray ($\overrightarrow{AD}$) after the pan (α) and tilt (β).

the mirror normal on the YZ plane (in both the PTU and LDV coordinate systems in Figure 4.3) as $\overrightarrow{AE}$ and $\overrightarrow{AF}$, respectively, we see that the angle $\angle BAF$ is α and the angle $\angle FAC$ is β. Now the camera optical axis is parallel to the mirror normal AC. Two additional angles are defined in the figure, the pan angle α' as the angle $\angle FAE$, and the tilt angle β' as the angle $\angle EAD$. If the master PTZ camera (mounted on top of the PTU) is tilted back by $-\beta$, then its optical axis will be parallel to AF. Therefore, by further panning the PTZ by the angle α' and tilting the PTZ by the angle β', its optical axis will be parallel to the reflected laser ray AD.

Defining a helping line GC parallel to AD, we can easily solve the additional pan (α') and tilt (β') based on triangulation. The detailed derivation is straightforward and is neglected here. As a result, the pan angle (α') is

$$\alpha' = \tan^{-1}\left(\frac{\tan}{\cos 2}\right) \tag{4.9}$$

and the tilt angle (β') is

$$\beta' = \sin^{-1}(\sin 2 * \cos) \tag{4.10}$$

4.7   EXPERIMENTAL RESULTS

In this section, we provide some results on distance measuring, surface selection, auto-aiming using laser-camera alignment, and surface focusing and listening using our multimodal sensory system.

4.7.1   DISTANCE MEASURING VALIDATION

This experiment is used to verify the accuracy of the calibration among sensory components. Therefore, the test is performed under controlled environments, inside a lab room (with distances up to 10 m) and in the corridor of a building (with distances up to 35 m). Note that the camera's focal lengths do not increase linearly with the change of the zoom levels. In order to perform accurate distance measurement on a large distance, we calibrated the focal lengths under different zoom factors.

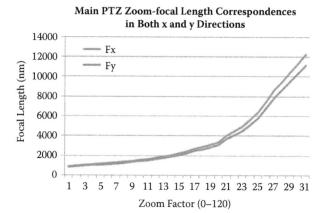

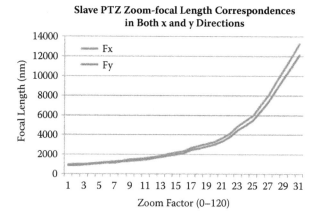

FIGURE 4.8   Calibrated focal lengths of the master PTZ camera and the slave PTZ camera under different zooms.

Figure 4.8 shows the focal lengths (in both x and y directions) of the main PTZ camera and the slave PTZ camera.

Next we verified the correctness of calibration parameters, especially with changes of the focal lengths of each camera. We used the same feature point on a checkerboard pattern at various distances. At each zoom level, the distance from the checkerboard to the platform was manually obtained as the ground truth, and then we used the calibrated parameters to estimate the distance at that zoom level. Figure 4.9 shows the comparison of the true and estimated distances under various zoom factors, which has an average relative error of 6%. The accuracy is sufficient for performing the adaptive focus of the LDV sensor.

4.7.2   SURFACE SELECTION

In this experiment (Figure 4.10), several interested target points are automatically selected in the segmented regions close to the human target in an image of the

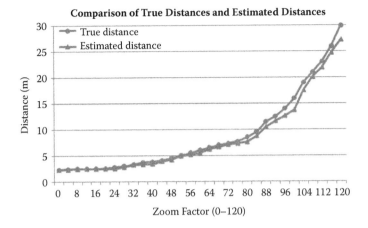

FIGURE 4.9   The comparison of true distances and estimated distances under various zoom factors.

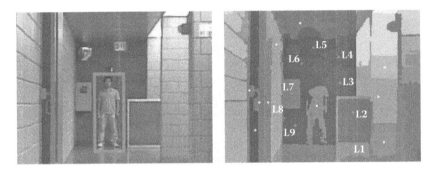

FIGURE 4.10   (See color insert.) The cropped (320 × 240) original image (under zoom factor 48) with a target (red rectangle) is shown on the left. On the right, interested target points close to the human target in the segmented regions are selected and labeled (as L1–L9). The distance of the target to the camera is about 31 m.

master PTZ. Note that a static target object such as a human or vehicle can be easily detected using histograms of oriented gradients (HOGs) (Dalal and Triggs, 2005) in the image. If a target is moving, then frame difference can be used to separate the target from the background surfaces. Then conventional color segmentation can be performed. The region centroid points close to the center of the target can be selected to point the laser.

4.7.3   AUTO-AIMING USING LASER-CAMERA ALIGNMENT

When an interested surface point is selected, the master camera is centered to that point; then the laser-camera alignment technique automatically aims the laser spot close to or right below the image center in focus using the calculated distance via stereo vision of the two PTZ cameras. Here we test our system in two environments: indoor (Figure 4.11) and outdoor (Figure 4.12).

FIGURE 4.11   The cropped images with circles showing the center of the original image include surfaces of a metal cake box, door handler, and extinguisher box (from left to right). The calculated distances are 8.9, 10.6, and 26.6 m with corresponding true distances at 9.0, 11.0, and 28 m.

FIGURE 4.12   Target surfaces are selected at a metal box under a tree, tape on a poster board, and a right turn sign. Their distances are 45.4, 45.5, and 53.3 m, respectively.

The indoor experiment is performed at the corridor at about 30 m. A metal box on a chair is placed on a fixed location at about 9 m. We manually selected three surface points: the points on the metal box (9 m), metal door handler (11 m), and extinguisher metal box (28 m). The laser spots can be clearly observed in the images that are close to the image centers with pixel errors of 2.3, 5.2, and 5.5 (from left to right in Figure 4.11).

The outdoor experiment is performed near a highway when the sensor platform has a standoff distance of about 60 m from the highway. Three sample surface targets to the side of the highway close to the sensor platform are selected, a metal box under a tree (45.4 m), a poster with tape (45.5 m), and a right turn sign (53.3 m). All images are zoomed so that both the image centers (circles) and the laser spots right below are visible. The average pixel difference between the laser spot and the image center for the three examples is 6 pixels.

4.7.4   SURFACE FOCUSING AND LISTENING

The experimental results related to the distance measuring, surface selection, and laser pointing for those labeled points (Figure 4.10) are presented in Table 4.1. The estimated camera distances (D) are listed in column 3 with the "ground truth" data (D^*) at column 6. The LDV distance (D_L) in column 4, the distance from the target point to the optical center of the LDV, is calculated based on the pan and tilt angles of the PTU. Based on that, the focus step (in the range of 0 to 3300) in column 5 is determined and the laser beam is focused in about 1 s for each point. For comparison,

TABLE 4.1
Surface Selection, Laser Pointing, and Focusing

L#	Surface	Measurements				Ground Truth		
		D (m)	D_L (m)	Step	Level	D* (m)	Step	Level
L1	Floor	26.56	27.14	2642	10	27.74	2581	11
L2	Chalkboard	28.30	28.88	2750	31	27.74	2764	22
L3	Wall	27.67	28.25	2732	12	30.63	2734	12
L4	Wall	28.62	29.20	2734	11	30.63	2734	12
L5	Wall	29.67	30.26	2786	12	30.63	2845	14
L6	Mirror	27.90	28.56	2758	12	30.63	2757	12
L7	Metal box	29.67	30.26	2745	118	30.32	2839	121
L8	Sidewall	21.10	21.60	2391	9	23.16	2410	10
L9	Wall	30.85	31.43	2410	11	30.63	2757	11

TABLE 4.2
Focus Positions and Signal Levels of Three Outdoor Surfaces

No.	Target	Distance (m)	Focus Position	Signal Level
001	Box under a tree	45.4	2890	285
002	Poster with tape	45.5	2890	116
003	Right turn sign	53.3	2904	14

the focus step using the full-range search takes 15 s, and is presented in column 8. The signal returning levels (0 to 512) in column 6 can be used to determine what the best point is among the candidates for audio acquisition. As a result, the metal box (L7) has the strongest signal return level, so it is selected for the voice detection. Note that all selected surfaces do not have retroreflective tape treatment.

The experiment results of the focus positions and signal levels of the outdoor surface targets (Figure 4.12) are shown in Table 4.2. According to the signal return levels at the last column, the surface of the metal box under a tree should be selected as the best listening target. In addition, the poster with tape is also a good listening surface with moderate signal level. Therefore, it can be used as a substitute for the first one with some signal-enhancing treatments, such as amplification, noise removal, and filtering. Unfortunately, the right turn sign does not provide sufficient signal returns.

4.8   CONCLUSIONS

In this chapter, we present a dual-PTZ-camera-based stereo vision system for improving the automation and time efficiency of LDV long-range remote hearing. The closed-loop adaptive sensing using the multimodal platform allows us to determine good surface points and quickly focus the laser beam based on target detection, surface point selection, distance measurements, and LDV signal return feedback. The integrated system greatly increases the performance of the LDV remote hearing, and therefore its feasibility for audiovisual surveillance and long-range inspection and detection of other applications. Experimental results show the capability and feasibility of our sensing system for long-range audio-video-range data acquisition.

ACKNOWLEDGMENTS

This work was supported by the U.S. Air Force Office of Scientific Research (AFOSR) under Award FA9550-08-1-0199 and the 2011 Air Force Summer Faculty Fellow Program (SFFP), by the Army Research Office (ARO) under DURIP Award W911NF-08-1-0531, by the National Science Foundation (NSF) under grant CNS-0551598, by the National Collegiate Inventors and Innovators Alliance (NCIIA) under an E-TEAM grant (6629-09), and by a PSC-CUNY research award. The work is also partially supported by NSF under award EFRI-1137172. We thank Dr. Jizhong Xiao at City College of New York and Dr. Yufu Qu at Beihang University, China, for their assistance in the prototyping of the hardware platform, and Dr. Rui Li for his discussions in the mathematical model for sensor alignment. An early and short version of this work was presented at the 2011 IEEE workshop on applications of computer vision (Wang et al., 2011b).

REFERENCES

Beal, M. J., Jojic, N., and Attias, H. 2003. A graphical model for audiovisual object tracking. *IEEE Trans. Pattern Anal. Machine Intell.*, 25, 828–836.

Blackmon, F. A., and Antonelli, L. T. 2006. Experimental detection and reception performance for uplink underwater acoustic communication using a remote, in-air, acousto-optic sensor. *IEEE J. Oceanic Eng.*, 31(1), 179–187.

Bouguet, J. Y. 2008. Camera calibration toolbox for Matlab. http://www.vision.caltech.edu/bouguetj/calib_doc/index.html

Clavel, C., Ehrette, T., and Richard, G. 2005. Events detection for an audio-based surveillance system. In *IEEE ICME '05*, Amsterdam, Netherlands, pp. 1306–1309

Cristani, M., Bicego, M., and Murino, V. 2007. Audio-visual event recognition in surveillance video sequences. *IEEE Trans. Multimedia*, 9(2), 257–267.

Dalal, N., and Triggs, B. 2005. Histogram of oriented gradient for human detection. In *Proceedings of IEEE Conference on Computer Vision and Pattern Recognition*, San Diego, California, pp. 886–893.

Dedeoglu, Y., Toreyin, B. U., Gudukbay, U., and Cetin, A. E. 2008. Surveillance using both video and audio. In *Multimodal processing and interaction: Audio, video, text*, ed. P. Maragos, A. Potamianos, and P. Gros, Springer LLC, pp. 143–156.

Khan, A. Z., Stanbridge, A. B., and Ewins, D. J. 1999. Detecting damage in vibrating structures with a scanning LDV. *Optics Lasers Eng.*, 32(6), 583–592.

Lai, P., et al. 2008. A robust feature selection method for noncontact biometrics based on laser Doppler vibrometry. In *IEEE Biometrics Symposium*, pp. 65–70.

Li, W., Liu, M., Zhu, Z., and Huang, T. S. 2006. LDV remote voice acquisition and enhancement. In *Proceedings of IEEE Conference on Pattern Recognition*, vol. 4, Hong Kong, China, pp. 262–265.

Li, X., Chen, G., Ji, Q., and Blasch, E. 2008. A non-cooperative long-range biometric system for maritime surveillance. In *Proceedings of IEEE Conference on Pattern Recognition*, Tampa, Florida, pp. 1–4.

Ometron. Ometron systems. http://www.imageautomation.com/ (accessed December 2010).

Polytec. Polytec laser vibrometer. http://www.polytec.com/ (accessed December 2010).

Qu, Y., Wang, T., and Zhu, Z. 2010. An active multimodal sensing platform for remote voice detection. In *IEEE/ASME International Conference on Advanced Intelligent Mechatronics (AIM 2010)*, Montreal, July 6–9, pp. 627–632.

Radhakrishnan, R., Divakaran, A., and Smaragdis, A. 2005. Audio analysis for surveillance applications. In *IEEE WASPAA '05*, New York, N.Y., pp. 158–161.

Scruby, C. B., and Drain, L. E. 1990. *Laser ultrasonics technologies and applications*. New York: Taylor & Francis.

Wang, T., Zhu, Z., and Taylor, C. 2011a. Multimodal temporal panorama for moving vehicle detection and reconstruction. Presented at IEEE ISM International Workshop on Video Panorama (IWVP), Dana Point, CA, December 5–7.

Wang, T., Li, R., Zhu, Z., and Qu, Y. 2011b. Active stereo vision for improving long range hearing using a laser Doppler vibrometer. In *IEEE Computer Society's Workshop on Applications of Computer Vision (WACV)*, Kona, Hawaii, January 5–6, pp. 564–569.

Zhu, Z., Li, W., Molina, E., and Wolberg, G. 2007. LDV sensing and processing for remote hearing in a multimodal surveillance system. In *Multimodal surveillance: Sensors, algorithms and systems*, ed. Z. Zhu and M. D. Huang. Artech House, 2007, pp. 59–88.

Zhu, Z., Li, W., and Wolberg, G. 2005. Integrating LDV audio and IR video for remote multimodal surveillance. Presented at OTCBVS '05, San Diego, California.

Zieger, C., Brutti, A., and Svaizer, P. 2009. Acoustic based surveillance system for intrusion detection. In *IEEE ICVSBS '09*, Genoa, Italy, pp. 314–319.

Zou, X., and Bhanu, B. 2005. Tracking humans using multimodal fusion. In *Proceedings of IEEE Conference on Computer Vision and Pattern Recognition*, San Diego, California, vol. 3, pp. 4–12.

5 High-Speed Fluorescence Imaging System for Freely Moving Animals

Joon Hyuk Park

CONTENTS

The study of neuronal activity in awake and freely moving animals represents a chance to examine neuronal function during natural behavior without the stress and suppression of activity inherent in the use of anesthetics and physical restraints. While electrophysiologic methods have produced profound insights into neuro-physiological events associated with sensory activation, motor tasks, sensory motor integration, and cognitive processes, these methods can only monitor a few dozen neurons simultaneously in an awake, behaving animal.

Optical probes can provide high spatial and temporal resolution, while being less invasive than traditional microelectrode methods. Scientific image sensors tailored for optical probes exist, but they are designed for fixed-animal experiments and are not suitable for behaving animals. Thus, a miniature camera, built around a custom-designed image sensor, is required to record neuronal activity in rodents with optical probes.

This chapter describes complementary metal-oxide-semiconductor (CMOS) image sensors specifically designed for a head-mountable imaging system for rodents. A small, self-contained imaging system with high sensitivity (>50 dB), speed (>500 frames/s), and large field of view (>2 mm²) is presented. The system consumes less than 100 mW of power and does not require active cooling. This work represents a first-generation, mobile scientific-grade physiology imaging camera.

5.1 INTRODUCTION TO CMOS IMAGE SENSORS

5.1.1 PHOTODIODES

Photodiodes are devices that convert incident light into current or voltage. Silicon is an ideal material for photodetectors in the visible wavelengths due to its band gap of 1.1 eV. Visible light between 450 and 650 nm has 2.75 to 1.9 eV of photon energy. Since these photons have an energy greater than the band gap of silicon, any incident photon in the visible wavelength will be absorbed in the silicon and produce charge. Also, these wavelengths are absorbed exponentially from the surface based on their energy, so blue light is mostly absorbed at the surface, while red light penetrates deeper into the silicon.

A photodiode is created by joining a P-type silicon with an N-type silicon to form a P-N junction. A depletion region is formed at the junction. The width of this depletion region can be controlled by changing the thicknesses of the outer P-layer, substrate N-layer, and bottom N+-layer, as well as the doping concentration and reverse bias voltage placed between the P- and N-layers. The width of the depletion region affects the spectral response of the photodetector—if the energy of the photon striking a photodetector is greater than the band gap energy of silicon, an electron within the crystal structure is pulled up into the conduction band. This leaves holes in their place in the valence band, creating an electron-hole pair where absorption occurs. The absorption can occur in either the depletion region, N-layer, or P-layer. The total current generated in the photodiode by incident photons is the addition of two diffusion currents (due to absorption in the P- or N-layer) and the drift current (due to absorption in the depletion region) (Figure 5.1).

PN photodiode-based pixels are most common in CMOS image sensors because they can be incorporated easily into a CMOS process. Any combination of different P- to N-layer junctions will create a photodiode, but the most common CMOS photodiode types are N-well/P-sub and N+/P-sub. Structures with N-well photodiodes improve the ratio of the number of charge carriers generated to the number of incident photons (quantum efficiency) due to a better charge collection from a greater depletion area (Tian et al., 2001b). There are various factors to consider when

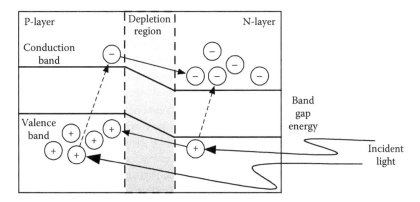

FIGURE 5.1   Generation of electrons and holes in a PN photodiode. When a photon with enough energy is absorbed in the N-layer, an electron is pulled into the conduction band, while a hole slowly diffuses into the depletion region, and then drifts into the P-layer by the electric field in the depletion region. When a photon is absorbed in the P-layer, the electron in the conduction band slowly diffuses toward the depletion region and then drifts to the N-layer, while the hole stays in the P-layer. Any electron and hole generated in the depletion region are immediately drifted to the N-layer and P-layer, respectively.

choosing a structure, as it can effect responsivity (sensitivity, given as watts per amp) at different wavelengths (due to shifting of the depletion region) and conversion gain (due to capacitance) (Perry, 1999). However, for both types of photodiodes, dark current due to surface generation and thermal noise associated with the photodiode reset must be addressed to increase the signal-to-noise ratio.

5.1.2 PINNED PHOTODIODE

A pinned diode structure has been introduced to the N-well/P-substrate photodiode to mitigate surface generation. The diode is created from a P+/N+/P-substrate, where the N+ region is pulled away from the silicon surface in order to reduce the surface defect noise (such as due to dark current) (Walden et al., 1972). As the voltage applied to the N+ layer is increased, the depletion regions of both P-N junctions grow toward each other. At a certain voltage, the pinned voltage Vpin, the depletion regions meet. The signal charge is generated and collected in this pinned diode region, isolated from the sense node and the readout circuitry by a transfer gate.

However, the maximum theoretical voltage swing is only $\frac{1}{2}$ Vdd. This is due to the fact that the pinned diode potential needs to be set approximately midway between the potential of the transfer gate when low and the potential of the sense node after reset. Also, a pinned photodiode requires two implants in the silicon: phosphorus, which defines the potential of the photodiode region, and boron, which pins the surface potential, and hence the diode potential, to ground. If the implants are not very accurately aligned, charge transfer problems will result. For example, a trap problem often develops when the p+ pinning implant encroaches under the transfer gate during high-temperature processing. This decreases the potential at the leading edge of the gate, resulting in a barrier for electron transfer. The fringing fields between the transfer gate and sense node region do not overcome the barrier because the potential difference between regions is very small. The CMOS trap problem described here can be corrected by offsetting (or angling) the pinning implant from the transfer gate slightly (less than 0.4 μm) (Janesick et al., 2002).

5.1.3 PHOTOTRANSISTOR

A phototransistor is a bipolar transistor, where the photons can reach the base-collector junction. The charge generated by photons in the base-collector junction is then injected into the base, and this current is amplified by the transistor's current gain β. This gain allows a phototransistor to produce a current output that is several times larger than a same area size photodiode due to the high gain of the transistor. Like the photodiode, there are a number of possible configurations for the phototransistor in a standard CMOS process, such as the vertical and lateral PNP phototransistors (Titus et al., 2011). However, since the photocurrent acts as base current into the transistor, phototransistors have low bandwidth, typically limited to hundreds of kHz. Additionally, the output of a phototransistor is linear for only three to four decades of illumination (since the DC gain of the phototransistor is a function of collector current, which in turn is determined by the base drive, the base drive may be in the form of a base drive current or incident light), compared to linearity of seven to nine decades of illumination for a photodiode (PerkinElmer, 2001). Lastly, the gain of the phototransistor can negatively impact performance. While the gain can provide higher quantum efficiency, at low-light levels, the emitter current of the phototransistor can be below 100 pA. When recombination within the base becomes a significant proportion of the emitter current, β becomes a decreasing function of the emitter current (Muller et al., 2002). Thus, the quantum efficiency (QE) of a phototransistor will be similar to that of a photodiode in low-light conditions, while β will also amplify the noise of the transistor. For these reasons, the phototransistor has limited applications.

5.1.4 AVALANCHE PHOTODIODE

The avalanche photodiode (APD) is the solid-state equivalent of a photomultiplier tube. Its main application is for detection of weak optical signals, such as single photon events. The APD exploits the impact ionization of carriers by photogenerated carriers under extremely high reverse bias. During its operation, carriers that trigger impact ionization and carriers that are generated by impact ionization continue to drift to cause more impact ionization events to occur. The result is a cascade of impact ionization events that produces an avalanche effect to amplify the photogenerated current.

An APD can also be operated in Geiger mode under reverse bias beyond the breakdown voltage. In this case, an electron-hole pair generated by a single photon will trigger the avalanche effect that generates a large current signal. Therefore, photon counting (like particle counting with a Geiger counter) can be achieved in Geiger mode. An APD that can operate in Geiger mode is also known as a single-photon avalanche diode (SPAD).

5.1.5 PIXEL DESIGNS

In a passive pixel sensor (PPS), the charge integrated on the photodiode is read out by a column-level charge amplifier (Figure 5.2). Similar to a DRAM, reading the photodiode is destructive. Since there are no active elements to drive the column lines, the charge transfer can take a long time. The main advantage of a PPS is that it has 1 transistor per pixel, allowing a small pixel or large fill factor. The main disadvantages are slow readout speed and increased noise—the temporal noise is directly proportional to the vertical resolution of the imager and indirectly proportional to the pixel area (Fujimori and Sodini, 2002).

An active pixel sensor (APS) uses intrapixel charge transfer along with an in-pixel amplifier (Fossum, 1993). One advantage of the active pixel over a passive pixel is the in-pixel amplification of the photogenerated charge inside the pixel, which suppresses the noise in the readout path. Also, the output of the photodiode is buffered by the in-pixel amplifier, and reading is nondestructive and much faster than with PPS. The nondestructive read, along with other on-chip circuitry, allows true correlated double sampling (CDS), low temporal noise operation, and fixed-pattern noise reduction.

A 4T pixel is a modification of APS. Besides the pinned photodiode, the pixel consists of four transistors (4T) that include a transfer gate (TX), reset transistor (RST), source follower (SF), and row-select (RS) transistor. The transfer gate separates the

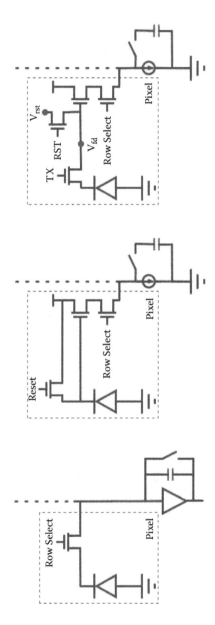

FIGURE 5.2 Different pixel types: a passive pixel sensor (left), 3T active pixel sensor (middle), and 4T active pixel sensor (right).

floating diffusion (FD) node from the photodiode node, which makes a true CDS readout possible, greatly reducing reset noise and fixed-pattern noise (FPN).

Compared to PPS, APS can be operated faster, with a higher signal-to-noise ratio (SNR). However, APS will have a larger pixel or lower fill factor than a PPS. The real bottleneck in terms of speed for the APS is the column readout time (as opposed to row transfer time for PPS).

5.1.6 READOUT SCHEMES

There are three main readout schemes employed in CMOS image sensors: pixel serial, column parallel, and pixel parallel. In pixel serial processing architecture, one pixel is chosen for readout by the row- and column-select signals at a time. Each pixel is processed sequentially. Without a global shutter, the integration time is shifted pixel by pixel.

In column parallel readout, pixels in a row are read out simultaneously and processed in parallel, stored in a line memory, and then read out sequentially. Without a global shutter, the integration time is shifted row by row. This architecture is the most popular scheme used in CMOS image sensors. Since all the pixels in the row are processed in parallel, the column processors can operate at lower frequencies than the pixel serial scheme, reducing overall power consumption.

In pixel parallel readout, a processor element (PE) is located in each pixel and image processing is performed in parallel for all pixels. Then, a compressed signal or only a signal of interest is read through the global processor. This scheme is used for very high-speed image processing applications (Kleinfelder et al., 2001), such as machine vision, where parallel processing is paramount. However, the pixel design becomes much more complex, resulting in a large pixel size or low fill factor.

5.1.7 SHUTTER TYPES

Two shutter types are employed to control the exposure time in CMOS image sensors. A rolling shutter resets each pixel sequentially in a pixel serial readout scheme, resulting in an integration time shift for every pixel. In column parallel readout, a rolling shutter results in an integration time shift for every row. While the rolling shutter improves readout speed, and thus frame rate, the shifted exposure times distort the images of moving objects. This can be mitigated with higher frame rates or by using a mechanical shutter.

A global shutter can fix the distortion problem, but requires a storage element in each pixel, increasing pixel size or lowering the fill factor. With a global shutter, the addressing and readout schemes become more complex, since the pixels need to be addressed globally, and then read out pixel sequential or column parallel. This also reduces frame rate since there can be no parallel operation within the imaging array during readout.

5.2 PERFORMANCE METRIC FOR IMAGE SENSORS

There are many properties of a pixel that can be characterized and define the performance of a pixel. These properties directly impact the image quality.

5.2.1   PIXEL CHARACTERISTIC

5.2.1.1   Responsivity

The responsivity of a photodetector is a measure of the conversion of the light power into electrical current. It is defined as the ratio of the photocurrent I_p to the incident light power P at a given wavelength:

$$R = \frac{I_P}{P}$$

The responsivity of a photodetector varies with the wavelength of the incident light, applied reverse bias, and temperature. Reverse bias slightly increases responsivity by improving the charge collection efficiency in the photodetector. Change in temperature can decrease or increase the band gap, which also impacts responsivity.

5.2.1.2   Dark Current

In an ideal photodetector, there is no current flow when no light is incident on the sensor. However, when a diode is reverse biased, a small leakage current flows. The impact of this leakage current on noise is more significant in newer processes with smaller pixel sizes. It originates from three sources: irregularities in the silicon lattice at the surface of the photodiode, currents generated as a result of the formation of a depletion region, and currents from the diffusion of charges in the bulk of silicon (Bogaart et al., 2009).

Whether or not a light is incident on the sensor, the dark current contributes to the current from drift and diffusion in the photodetector. Dark current is temperature and voltage dependent: increasing temperature or increasing reverse bias increases the dark current.

5.2.1.3   Quantum Efficiency

Quantum efficiency is defined as the number of electrons produced by incident photons on the pixel. It is a measure of how responsive the sensor is to different wavelengths of light and is related to responsivity by

$$QE_\lambda = \frac{R_\lambda}{q} \times \frac{hc}{\lambda} = \frac{1240 \times R_\lambda}{\lambda}$$

where R_λ is the responsivity at wavelength λ.

5.2.1.4   Full Well Capacity

Full well capacity is a measure of the largest charge, $Q_{D,max}$, that can be stored in the photodiode. This is a physical limit on the highest detectable signal of the sensor. It is given by

$$Q_{D,max} = C_D V_{max}$$

where C_D is the photodiode capacitance and V_{max} is maximum voltage swing, defined by the reset level and limits of the readout circuitry.

5.2.1.5 Dynamic Range

The well capacity impacts another metric, the dynamic range. The dynamic range, measured in decibels, is the ratio between the smallest and largest possible values read out from the sensor. It is defined as the signal value at full well capacity minus the readout noise value. It determines how well a sensor can measure an accurate signal at low-light intensities all the way up until it reaches full well capacity. A higher dynamic range will allow the sensor to operate in a wider range of lighting conditions.

$$DR = 20 \log_{10} \frac{V_{max}}{V_{noise,readout}}$$

5.2.1.6 Conversion Gain

The conversion gain of a sensor is the relationship between voltage swing and optical input power, and is a function of the photodetector's responsivity, amplifier gain (A_v), and input impedance. It expresses how much voltage change is obtained by the absorption of one charge:

$$CG = A_v \times \frac{q}{C_D}$$

5.2.1.7 Fill Factor

Fill factor (FF) is the ratio of the photosensitive area inside a pixel, A_D, to the pixel area, A_{pixel}:

$$FF = \frac{A_D}{A_{pixel}} \times 100\%$$

Thus, the fill factor will decrease as more electronics are placed within a pixel. Since CMOS pixels usually have three or more transistors (although some have 1.5 transistors per pixel (Kasano et al., 2006)) for basic operation, the fill factor will always be less than 100% when front illuminated. A lower fill factor will require a longer exposure time, and can make low-light photography impossible.

5.2.2 SENSOR NOISE

The noise of a sensor is important when detecting low-level radiation because it dictates the minimum detectable signal. There are three main types of noise associated with CMOS image sensors, which are discussed below. Other noise sources exist, such as analog-to-digital converter (ADC) noise, but they are either negligible or difficult to control or measure, so they are not included in our sensor tests.

5.2.2.1 Temporal Noise

Temporal noise is a time-dependent fluctuation in the signal. It is harder to compensate than fixed-pattern noise. The components of the temporal noise are dark-current shot noise, photon shot noise, reset noise, and read noise.

5.2.2.1.1   Dark-Current Shot Noise

Dark-current shot noise is a pixel shot noise associated with the photodiode leakage current I_{dark}, and is largely dependent on exposure time. It is given (in terms of volts) by

$$Noise = \frac{q}{C_D} \times \sqrt{\frac{I_{dark}\tau}{q}}$$

where C_D is the photodiode capacitance and τ is the integration time.

5.2.2.1.2   Photon Shot Noise

Photon shot noise is the majority component of overall noise at medium to high light levels. This noise derives from the number of photons hitting a pixel, and the thermally generated charges within the pixel fluctuating according to Bose-Einstein statistics:

$$\sigma_{shot}(P_I)^2 = P_I \frac{e^{\frac{hc}{\lambda kT}}}{e^{\frac{hc}{\lambda kT}} - 1}$$

where $\sigma_{shot}(P_I)^2$ is the photon shot noise variance, P_I is the average number of incident photons, h is Planck's constant, λ is the photon wavelength, k is Boltzmann's constant, c is the speed of light, and T is the temperature.

At room temperatures, and with photons in the visible spectrum,

$$\frac{hc}{\lambda} \gg kT \ ,$$

so $\sigma_{shot}(P_I) = \sqrt{P_I}$.

Thus, the photon shot noise has a fixed theoretical limit, and the only way to reduce it with respect to the overall signal (increase SNR) is to increase the full well capacity.

5.2.2.1.3   Reset Noise

The reset noise, also known as the kT/C noise, originates from random fluctuations in the voltages read from the capacitors in the photodiode. This fluctuation is due to the Johnson noise associated with the reset metal-oxide-semiconductor field-effect transistor (MOSFET) channel resistance. The noise voltage associated with the switch resistance (Rs) of the reset MOSFET is

$$v_{n,reset} = \sqrt{4kTR_s\Delta f}$$

where k is Boltzmann's constant, T is temperature, and Δf is the electrical bandwidth. The switch resistance and photodiode capacitance limits the equivalent bandwidth:

$$\Delta f = \frac{\pi}{2} \times \frac{1}{2\pi R_s C_D}$$

Combining the two equations:

$$v_{n,reset} = \sqrt{\frac{kT}{C_D}}$$

The signal integrated on a photodiode is measured relative to its reset level, so the reset noise can be a major contributor of noise. Reset schemes must be devised to reduce or eliminate this pixel-level noise.

5.2.2.1.4   Read Noise

Due to the source follower in APS, thermal and flicker ($1/f$) noise are present in each pixel. Until recently, reset noise has been dominant at the pixel level, but the use of correlated double sampling have mitigated this problem. Thus, thermal and flicker noise make up a large portion of read noise at the pixel level. Thermal noise is a function of dark current and increases with temperature and integration time.

Flicker noise in the pixel originates during two time frames. During integration, the $1/f$ noise is mostly generated by the (reverse-biased) photodiode due to surface recombination of carriers and fluctuation in bulk carrier mobility. However, it is a function of the dark current and is a minor component of noise compared to thermal noise. During readout, the $1/f$ noise originates from the source-follower transistor and the row-select transistor. The $1/f$ noise is typically higher than thermal noise for the follower transistor and much lower than thermal noise for the row-select transistor (Tian and El Gamal, 2001).

Many CMOS image sensors employ other circuitry, and each component contributes to the read noise. Since the column amplifier samples both the pixel reset and signal level, it contributes kT/C noise associated with the sampling process, along with thermal and $1/f$ noise of the column amplifier MOS devices. However, the thermal and flicker noise sources of the column amplifier are generally negligible when compared to the sampling operation kT/C noise (Hewlett-Packard, 1998). Since the two sampling operations are uncorrelated, the column amplifier kT/C noise is given by

$$v_{n,sample} = \sqrt{\frac{2kT}{C_{column}}}$$

Modern CMOS imagers typically contain an ADC, and the primary source of noise in the ADC is the quantization noise. An ideal ADC's quantization noise is given by

$$v_{n,ADC} = \frac{LSB}{\sqrt{12}} = 0.288 \times LSB$$

ADC noise will typically exceed 0.288 least significant bit (LSB) due to other noise sources, like thermal, amplifier, and switching noise.

5.2.2.2   Fixed-Pattern Noise

Fixed-pattern noise is a nontemporal variation in pixel values. The mismatch between values arises from multiple sources in the readout path: the minor differences or imperfections in the electrical components in each pixel, variations in column amplifiers, and random mismatches in ADC components. These mismatches produce a spatially static noise pattern on the sensor. FPN is not dependent on the number of interacting photons and is only a dominant noise source at low-light levels. It can be either coherent or noncoherent.

Noncoherent, dark-current FPN is due to mismatches in pixel photodiode leakage currents, and it typically dominates the noncoherent component of FPN, especially with long exposure times. Low-leakage photodiodes can reduce dark FPN. Another method is to perform dark-frame subtraction, at the cost of increased readout time.

Coherent, row- and column-wise FPN is due to mismatches in multiple signal paths and uncorrelated, row-wise operations of the image sensor. Coherent FPN offset components can be mostly eliminated by reference frame subtraction. Gain mismatches are more difficult to remove since gain correction is required, which can be time- and hardware-intensive.

5.2.3 Methods to Improve Performance

5.2.3.1 Correlated Double Sampling

Reset noise is dominant at low signal levels, since it is not dependent on the number of incident photons. It is the dominant source of noise in low-light conditions, but can be suppressed with correlated double sampling. CDS involves sampling the pixel value twice: once right after rest and then with the signal present. By taking the difference between these two values, the offset and reset noise are eliminated (White et al., 1974). It is fairly easy to implement in a pinned photodiode, but requires a storage capacitor in a 3T APS for true CDS (Anaxagoras et al., 2010).

Also, since CDS is essentially performing dark-frame subtraction, it can reduce dark FPN. Thus, when CDS is used, achieving a high SNR with the sensor is limited by shot noise and readout noise. However, CDS can result in reduced fill factor for 3T APS due to the requirement of additional components inside a pixel.

Furthermore, the minor components of readout noise can be further reduced with little effort. Flicker noise can be mostly eliminated through rapid double sampling of the pixel. Thermal noise in MOS devices can be suppressed by limiting the bandwidth of in-pixel amplifiers. The bandwidth is naturally limited in high-resolution sensors, due to large capacitive loads on the pixel amplifiers (Figure 5.3).

5.2.3.2 Microlensing

As pixel sizes get smaller, the fill factor is reduced. The lower fill factor in turn requires increased exposure times. If the fill factor is too low, a flash is necessary for indoor photography. To mitigate this problem, a microlens can be added to each pixel (Ishihara and Tanigaki, 1983). The microlens gathers light from the insensitive areas of the pixel and focuses it down to the active area.

The microlensing process first planarizes the surface of the color filter layer with a transparent resin. Then, the microlens resin layer is spin-coated onto this planarization layer. Lastly, photolithographic patterning is applied to the resin layer and is shaped into a dome-like microlens by wafer baking.

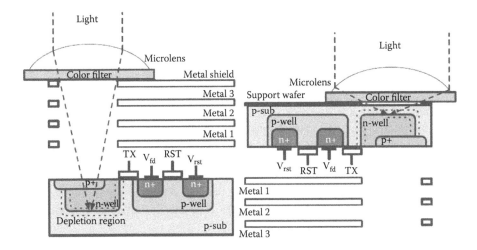

FIGURE 5.3 (See color insert.) Comparison of different pixel designs: front-side illumination (left) and backside illumination (right). For both, microlenses are shown. Transistors are not drawn to scale.

5.2.3.3 Backside Illumination

In a typical front-side-illuminated (FSI) CMOS image sensor, the photodiode element is at the bottom of all layers, including the metal layers. The metal layers can reflect some of the light, and the angle of the light is also limited by the depth of the layer.

It is possible to illuminate from the backside of the sensor by flipping the silicon wafer during manufacturing and then thinning the backside enough so that the incident light can strike the photosensitive layer before all other layers (Jerram et al., 2010). With a backside-illuminated (BSI) sensor the QE can improve to more than 90%. However, cross talk, which causes noise, dark current, and color mixing, can be introduced (Rao et al., 2011). Also, thinning the wafer also poses wire-bonding problems, as the wafer can become too thin, and becomes prone to damage. Lastly, microlenses on the FSI pixel can offer a QE close to that of a BSI pixel (Jerram et al., 2010) in smaller fabrication processes.

5.3 MOBILE CORTICAL IMAGING

Recent advancements in optical probes and CMOS image sensors make it possible to build a mobile imaging system capable of monitoring cortical activity over a wide area in behaving rodents. Currently available imaging systems are shown to be unable to record the activity in the neurons of awake and freely moving experiments. A set of design requirements are laid out below for a mobile imaging system to enable such recordings.

5.3.1 Optical Probes

Optical detection of rapid changes in membrane potential was first reported in 1968 (Cohen et al., 1968; Tasaki et al., 1968). There are two types of optical probes:

calcium sensitive and voltage sensitive. Calcium probes monitor the intracellular calcium level, and voltage probes monitor the membrane potential.

5.3.1.1 Voltage-Sensitive Dyes

A voltage-sensitive dye (VSD) is an organic molecule with a hydrophobic portion that sticks to the membrane and a charged chromophore that prevents entry to the cell interior. Thus, the dye only binds to the external surface of membranes and does not interrupt a cell's normal function. When bound to neuronal membranes, VSDs have a high absorption coefficient and a high quantum efficiency for fluorescence. When excited with the appropriate wavelength, VSDs emit a fluorescent light that is a function of the changes in membrane potential with high temporal resolution (Petersen et al., 2003). While pharmacological side effects and phototoxicity have been limitations of VSDs, newer dyes have mostly mitigated this problem (Shoham et al., 1999).

In voltage-sensitive dye imaging (VSDI) of the cortex, a single pixel typically contains a blurred image of various neuronal components (dendrites, axons, and somata of a population of neurons). The fluorescent signal from VSDs is linearly related to the stained membrane area. A large amount of the signal originates from cortical dendrites and nonmyelinated axons rather than cell bodies. Since dendrites of cortical cells are far more confined than the axons, most of this signal originates from the dendrites of neighboring cortical cells, reflecting dendritic activity. Thus, a signal in a particular site does not imply that cortical neurons at that site are generating action potentials. Even so, with techniques such as differential imaging (Blasdel and Salama, 1986), VSDI provides a high-resolution map of where spiking occurs in the cortex.

VSDI is an excellent technique to study the dynamics of cortical processing at the neuronal population level, and provides higher-level information about cortical neuronal processes than traditional electrophysiology methods (Davis et al., 2011). Such processes include propagating sub- and suprathreshold waves and state changes, electrical events in neuron classes generally inaccessible to electrodes, and glial responses.

5.3.1.2 Calcium-Sensitive Dyes

A calcium-sensitive dye (CSD) enters through the cell membrane and works on the inside of the cell. The change in fluorescence of the dye depends on the intracellular calcium level. Such dyes allow individual cell measurement and have a large signal, but loading the dye is more difficult than VSD. Voltage dyes bind to cell membranes and change its fluorescence, depending on the membrane potentials of the cell. They measure the population average, have a smaller signal, and are easier to stain than CSD.

The binding of calcium causes large changes in ultraviolet absorption and fluorescence. When calcium is bound, it fluoresces at different wavelengths than when it is not bound. Thus, by exciting the cell at two different wavelengths, the ratio of the fluorescence intensities at the two wavelengths gives the ratio of the concentrations of bound to free calcium. This is useful because Ca^{2+} is an intracellular signal responsible for controlling other cellular processes (Berridge et al., 2000).

Neurons have several Ca^{2+}-sensitive processes located in different regions. Action potentials in neurons cause a transient rise in intracellular Ca^{2+}. However, since calcium is a secondary indicator of action potentials, CSDs are slower to respond than VSDs (Baker et al., 2005). Also, CSDs act as buffers of calcium, so the calcium signal in cells loaded with the dye may last longer than in cells without the dye (Neher, 2000). Thus, the measure of calcium concentration in a neuron provided by CSDs must be interpreted with care.

5.3.1.3 Advantages over Other Methods

A traditional method to record neuronal activity in awake, unrestrained, and mobile animals at high speed requires electrode-based systems. While these electrophysiologic methods have produced profound insights into neurophysiological events associated with sensory activation, motor tasks, sensory motor integration, and cognitive processes, these methods can only monitor a few dozen neurons simultaneously in an awake behaving animal. The study of identified neurons or studies that require the absolute continuity of cell identity during a recording session cannot be performed with single-unit electrophysiology methods. Electroencephalograms and local field potentials suffer from point source localization problems due to the three-dimensional origins of activity and the complex structure of the conducting element, the brain tissue. Thus, the source of an electroencephalogram event cannot be accurately localized within the three-dimensional space of the brain even with a large number of scalp or penetrating electrodes (Figure 5.4).

Optical probes are a less invasive alternative/supplement to currently available methods. They provide high-level information of a neuronal population with high temporal and spatial resolution. VSDI is preferred over calcium sensitive dye imaging (CSDI). Also, with genetically encoded dyes, there is no need to load the optical sensors onto cells (Siegel and Isacoff, 1997). Instead, the genes encoding for these dyes can be transfected to cell lines. It is possible to have transgenic animals expressing the dye in all cells or only in certain cellular subtypes.

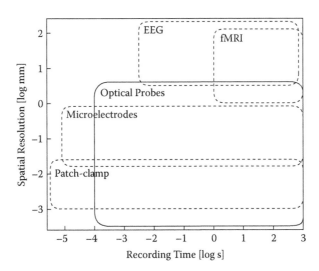

FIGURE 5.4   Comparison of VSDI/CSDI with other recording techniques.

5.3.2 NEED FOR MOBILE SYSTEMS

Functional magnetic resonance imaging and positron emission tomography allow three-dimensional recording of brain metabolism, but these techniques are indirect measures of neuronal activity, have low temporal and spatial resolution, and require head fixation. Conversely, implantable microelectrode arrays allow high-speed recording of neuronal activity in awake, unrestrained, and mobile animals. However, only a few dozen neurons in a small area can simultaneously be recorded in an awake, behaving animal using microelectrode arrays.

Currently, there are no optical recording methods suited for wide-field, high-resolution, high-speed imaging of voltage-sensitive sensors in freely moving animals. Fiber-optic-connected laser-scanning imaging methods (Flusberg et al., 2008; Helmchen et al., 2001) have produced high-resolution images of neurons in animals. However, these approaches currently lack the sensitivity, speed, and field of view to record the rapid (500 Hz) and relatively small fluorescence intensity changes produced by optical probes across a functionally relevant domain of cortex (i.e., visual cortex > 2 mm^2). A bundled optical fiber has been used to perform wide-field voltage-sensitive dye imaging in a freely moving rat (Ferezou et al., 2006), and the use of this device has produced impressive results about the activity in a barrel cortex during resting and active whisking (Ferezou et al., 2007). However, imaging with a fiber in a more complex experiment involving extensive movement is limited due to the fiber stiffness, size, and requirement of a commutator.

Thus, there is a need for an autonomous (untethered) system that allows optical recording of rapid electrical events over a large cortical area in freely moving animals. A small, self-contained imaging system was recently published (Ghosh et al., 2011). While compact, this design does not achieve the majority of our design requirements. The image sensor in that system can only image at 45 fps. The microscope body is made of black PEEK, which is autofluorescent with blue or green illumination. Also, the light source is neither stabilized nor bright enough for voltage probe imaging. Finally, the field of view of the microscope is only 0.6×0.8 mm. This instrument is adequate for highly fluorescent processes with large ΔF/F changes (i.e., small dye-saturated vascular perfusion studies or small-molecule calcium dyes), but is not suited for mobile applications studying physiologically relevant changes (i.e., >200 fps) in voltage (0.2–2% ΔF/F) over large areas of cortex. Another similar system for reflected light imaging in rats was published recently (Murari et al., 2010), but it also lacks the required SNR and speed for the study of neuronal electrical activity.

Commercial scientific-grade image sensors such as NeuroCCD (RedShirtImaging, Decatur, GA) and dicam (PCO, Kelheim, Germany) typically use charge-coupled devices (CCDs) due to lower read noise (80 dB peak SNR for NeuroCCD and 69.3 dB peak SNR for dicam) and higher quantum efficiency (80% for NeuroCCD and 50% for dicam) than CMOS sensors (Tian et al., 2001a). However, CCDs consume tens of watts (51 W for dicam) of power and the imaging systems have active cooling components weighing several pounds (8 kg for dicam). The main advantage of CMOS is the integration of peripheral circuitry at the pixel, column, or chip level. This enables on-chip image processing while consuming only milliwatts of power with passive cooling. While scientific-grade CMOS imagers such as MiCam Ultima

(Brainvision, Tokyo, Japan) are approaching the quality of CCD imagers, power and weight are still issues that need to be addressed. Commercial CMOS imagers are mainly designed for cell phones or professional photography, and are energy-intensive (>100 mW), slow (usually 30 fps), designed for color imaging, and do not have a high enough SNR to detect small changes in intensity. Cell phone CMOS sensors are small in size (about 3×3 mm) and consume tens of milliwatts of power, but have a SNR of less than 40 dB and are limited to 60 fps. Professional photography CMOS sensors have higher, yet inadequate SNRs of less than 50 dB and require too much power. Also, both of these classes of imagers have higher resolution (>1 Mpixels) than necessary for biological applications. Binning would have to be implemented to collect enough photons in typical biological conditions to increase the SNR, which reduces the overall resolution. A custom pixel array design based on the image sensor fabrication technology of professional photography CMOS sensors is preferred, but this technology is highly confidential and biological imaging does not currently have a large enough market to created interest from large CMOS image sensor companies.

5.4 IMAGING SYSTEM DESIGN

Voltage-sensitive dye imaging systems currently used for fixed-animal experiments cannot readily be adapted for use on freely moving animals due to the strict requirements of weight (<10 g) and power (<100 mW). A miniaturized implantable VSDI sensor has many design challenges. The SNR must be high, as the signal of interest is around 0.2–2% of the background illumination. Thus, a large photon collection area per pixel must be paired with a 10- to 12-bit ADC. A high bandwidth (>500 fps) is needed to sample fast neurophysiologic events in the order of milliseconds. Also, there are a number of dynamic noise sources that degrade the signal. A custom-designed miniaturized head-mounted imaging system that addresses all of these issues is required.

5.4.1 MICROSCOPE DESIGN

Since the weight of a rat is around 300 g, the complete imaging system must be weigh around 10 g for the animal to move around freely. A miniature microscope was built with Delrin, a strong, lightweight plastic. The light source was provided by a Philips Luxeon Rebel light-emitting diode (LED) with a custom-designed heat sink. The necessary optics (collimator, emission/excitation filters, and dichroic) used were all commercially available parts. The image sensors described below were interfaced with the microscope, and more information on the microscope design can be found in Osman et al. (2012).

5.4.2 A CMOS IMAGE SENSOR IN 0.5 µM PROCESS

A 3×3 mm image sensor was fabricated in AMI's 0.5 µm bulk CMOS. It has a 32×32 pixel array with a global shutter. For sizing the photodiode, the emission of the target voltage-sensitive probe was assumed to have a wavelength of 650 nm (di-8-ANEPPS has a peak emission wavelength of 605 nm and RH1691 peaks at 665 nm).

5.4.2.1   Design Methodology

The expected intensity of the photons hitting the image sensor during a typical VSDI experiment was measured with a calibrated photodiode, and was estimated to be around 1.59 µW, or about 133 pA/pixel. For the size of the photodiode, C_D is calculated to be 1.18 pF, so a 0.1% change near saturation would be 113 mV/s. Assuming a voltage swing of 1V, this would require an ADC of >13 bits.

Since the photodiode should have a capacitance of 1.18 pF for the designed size, reset noise can be estimated to be 59 µV. Also, the sensor will be operating with a global shutter, so a sampling capacitor in the pixel also must be sufficiently large to be a minor contributor to the kT/C noise. Charge injection will also add noise during sampling, so dummy switches need to be implemented. Assuming a 10 nA/cm² dark current for this particular process, for a photodiode of 75×35 µm, this would be 0.26 pA, or about 0.2% of the total signal. This static current can be negated with dark-frame subtraction, but the dark-current shot noise may pose an issue if the dark current is higher than assumed.

Column and row scan shift registers address each pixel during readout. A global buffer based on a cascode operational transconductance amplifier (OTA) design (Baker, 2010) is used to transfer the voltage value to an off-chip ADC. The pixel uses a modified 3T design composed of a P-type metal-oxide semiconductor (PMOS) reset transistor and a N-type metal-oxide semiconductor (NMOS) follower. The storage capacitor is a metal-oxide semiconductor capacitor (MOSCAP) sized for 788 fF, and allows the implementation of a global shutter. A dummy switch (controlled by s1N) is used to reduce clock feedthrough and charge injection. The pixel value can either be stored in a capacitor or read out through a line-out. During readout, the integrated voltage of the photodiode can be stored in the storage capacitor by turning on s1 (shutter). Once integration is finished, s1 can be turned off to isolate the storage capacitor from the photodiode. The stored voltage value is then read out by the circuit (Figure 5.5).

5.4.2.2   Sensor Characterization

The sensitivity of the pixel was calculated to be 0.62 V/lx·s by measuring the voltage integration at 50 lx with the sensor running at 500 fps. Since the full voltage swing of the pixel is 1.1 V, the full well capacity is 2.1 Me⁻, and the conversion gain is 0.52 µV/e⁻. To test the SNR of the image sensor, the photodiode in every pixel was integrated for 709 µs at different light intensities. The light was provided by a Luxeon Rebel LED controlled with a custom-built feedback-controlled drive circuit (an Arduino microcontroller and a National Instruments DAQ board). A set of 10,000 consecutive frames at each light intensity was collected, and the voltage drop of the photodiode was divided by the standard deviation of the values of the photodiode to calculate the SNR. The photodiode saturates at around 340 lux at 500 fps and has a SNR of 61 dB. The SNR needs to be between 55 and 72 dB for VSDI applications.

The quantum efficiency of the photodiode was measured from 250 to 1000 nm. The peak quantum efficiency of the photodiode was about 25%. This is due to the fact that the CMOS process used is not optimized for optical imaging applications. With a fill factor of 50%, the operational quantum efficiency of each pixel is 12.5%.

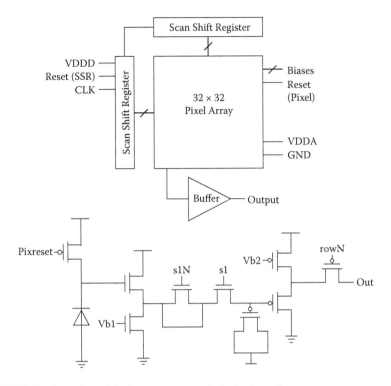

FIGURE 5.5   Overview of the image sensor and pixel schematic.

The pixel reset noise, directly related to the capacitance of the photodiode, is 115 μV. The voltage drop due to dark current has a slope of 0.14 V/s, so the dark current is 162 fA, or 6.4 nA/cm². At 500 fps, the dark shot noise is 3.8 μV. The board noise is calculated by connecting a very stable DC voltage to the output pin of the image sensor (input pin of the ADC) on the board and collecting 12,000 frames of all 32 × 32 pixels. The ADC was running at 1 million samples per second (MSPS), the highest achievable with this specific model (Analog Devices AD7980). The standard deviation of all pixels in the 12,000 frames was 138 μV. The noise due to all other components on the chip is calculated by running the image sensor with a very short integration time (approximately 1 μs) in the dark. Under these conditions, the total noise of the circuit was 300 μV. The only noise sources to take into consideration in such a short integration time in the dark are reset, board, and readout. Thus, the read noise is 260 μV. The FPN was calculated by measuring the standard deviation of each of the 32 × 32 pixels over 10,000 frames and then taking the mean of the resulting 1024 standard deviation values. The image sensor has a pixel FPN of 1.5%, a column FPN of 1.1% and a row FPN of 0.82%.

5.4.2.3   Miniaturization

The image sensor was placed in a miniature printed circuit board (PCB), measuring about 20 × 20 mm. The PCB contains an ADC, DAC, resistor-network voltage dividers, and bypass capacitors. The housing for the PCB weighs 3.4 g and measures 22.25 × 22.25 mm (Figure 5.6).

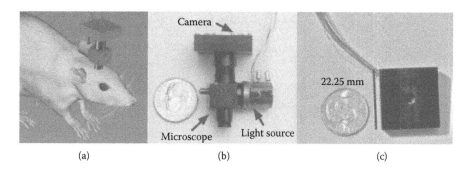

FIGURE 5.6 (See color insert.) (a) Proposed implementation of the image sensor in a self-contained system for imaging fluorescent activity reporters in freely moving rodents. (b) The camera described in this chapter attached to a head-mountable microscope system; umbilical cable not shown. The microscope contains an excitation filter, emission filter, dichroic, and lenses (objective, condenser, relay). The light source contains a LED and heat sink. (c) Close-up of the camera housing and umbilical cable.

5.4.2.4 Rat Olfactory Bulb Recordings

A CSD, Oregon Green BAPTA 488 Dextran 10 kD, was loaded onto rat olfactory receptor neurons 7 days prior to imaging using a procedure similar to that developed for mice (Verhagen et al., 2007). All animal experiments were approved by our Institutional Animal Care and Use Committee in compliance with all Public Health Service Guidelines. Just prior to imaging, an optical window was installed over the dorsal surface of each olfactory bulb by thinning the overlying bone to 100 μm thickness and coating the thinned bone with ethyl 2-cyanoacrylate glue to increase its refractive index. Odorants were diluted from saturated vapor of pure liquid chemicals. Linearity and stability of the olfactometer were verified with a photoionization detector. An odorant delivery tube (20 mm ID, 40 mm long), positioned 6 mm from the animal's nose, was connected to a three-way valve-controlled vacuum. Switching this vacuum away from the tube ensured rapid onset and offset of the odorant. After creating the optical window, a double tracheotomy was performed to control sniffing. To draw odorants into the nasal cavity, a Teflon tube was inserted tightly into the nasopharynx. This tube communicated intermittently (250 ms inhalation at a 2 s interval) with a vacuum via a PC-controlled three-way valve. A cyan LED was the light source. The Neuroplex software initiated a trial by triggering the onset of the light source, resetting the sniff cycle, and triggering the olfactometer. Imaging started 0.5 s after the trigger, and an odorant was presented from 4 to 9.5 s. Running at 25 fps, the change in intensity detected by our system with a 4% concentration of 2-hexanone delivered orthonasally was around 3.1%. This is comparable to the data collected with NeuroCCD-SM256, a commercial camera (Figure 5.7).

5.4.2.5 Mouse Somatosensory Cortex during Whisker Deflection

A 30 g mouse was anesthetized with urethane (0.6 ml, 100 mg/ml, IP). The depth of anesthesia was monitored by tail pinch and heart rate. Body temperature was

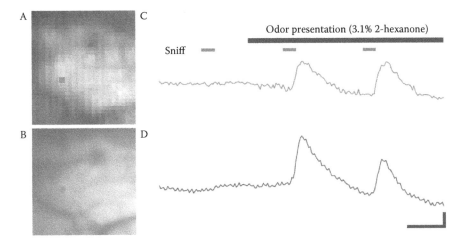

FIGURE 5.7   (A) Image of Oregon Green BAPTA–loaded olfactory nerve terminals in the rat olfactory bulb taken with the sensor described in this chapter. The black square indicates the location of a 2 × 2 pixel subregion averaged to produce the trace in C. (B) The same field of view captured with a NeuroCCD-SM256 camera (RedShirtImaging, LLC). The black square indicates an 8 × 8 pixel subregion averaged to produce the trace in D. (C) Calcium transients recorded from olfactory nerve terminals within glomeruli of the rat olfactory bulb following presentation of an odorant (3.1% 2-hexanone) recorded with our sensor. (D) Approximately same region of interest (ROI) imaged with the NeuroCCD-SM256 camera. Both cameras produce similar-sized signals with comparable noise characteristics. Some of the repetitive noise in panel D arose from metabolic (i.e., blood flow from heart pumping) sources. Both traces are single presentations and unfiltered. Vertical and horizontal scale bars represent 1% ΔF/F and 1 s, respectively.

maintained at 37°C by using a heating pad and rectal thermometer. Additional anesthesia was given as needed (0.1 ml, IP). A dye well/head holder was affixed to the exposed skull using cynoacrylic glue. A craniotomy (5 mm diameter) was performed over barrel cortex (2.1 mm lateral and 2 mm caudal to bregma) using a Gesswein drill with a 0.5 round head bit. The dura was left intact and a solution of RH1692 (100 μl, 0.1 mg/ml of ACSF) was applied for 1.5 h with stirring every 10–15 min to prevent the CSF from forming a barrier between the dye and the brain. A pressurized air puff (50 ms, Picospritzer) was used to deflect the whiskers contralateral to the side of imaging. The barrel cortex was imaged using a tandem-lens epifluorescence macroscope equipped with a 150 W xenon arc lamp (Opti-Quip, Highland Mills, New York) (Figure 5.8).

5.4.3   CMOS IMAGE SENSOR IN 0.35 OPTO

A second image sensor was designed in Austria Microsystems' 0.35 μm OPTO process. While this process does not offer a pinned photodiode, the higher well capacity offered by a N-well/p-sub photodiode is also desirable for VSDI. The OPTO process also offers low dark current and an ARC to improve the QE.

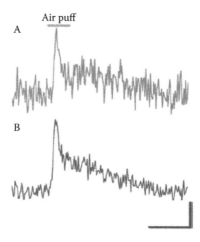

FIGURE 5.8 Data captured from a voltage-sensitive dye (RH1692)–stained barrel cortex following whisker deflection by a 50 ms air puff. (A) Twenty-trial average of a 138 × 138 μm area of cortex with our imaging system (2 × 2 pixels). (B) Twenty-trial average of the same area of the same experiment using a RedShirtImaging NeuroCCD (4 × 4 pixels). Data were collected at 500 Hz for both systems. The vertical scale bar represents 0.5% ΔF/F, and the horizontal scale bar represents 0.1 s.

5.4.3.1 Design Methodology

The new image sensor is designed within a 4 × 4 mm die area. The resolution of the previous image sensor (32 × 32) was acceptable, but a 100 × 100 resolution is implemented in this design. Also, biasing, state machine, and ADC circuits are implemented on-chip to reduce the overall size of the imaging system.

A 30 × 30 μm pixel with a 20 × 20 μm photodiode size was chosen in a trade-off between resolution, well capacity, and available chip area. A photodiode of that size would result in a capacitance of about 32 fF, according to the process data sheet. Assuming a 1 V swing, that would be 200,000 e– well capacity. By comparison, NeuroCCD-SMQ has a well capacity of 215,000 e–. Assuming ideal conditions, the shot-noise-limited SNR would be 53 dB. Thus, near saturation, a 0.2% change in intensity can be detected. The pixel is designed to operate with either a global shutter or a rolling shutter. Thus, a storage capacitor of 140 fF is implemented as an nMOSCAP. A MOSCAP is not as linear as a metal-insulator metal (MIM) or polymer capacitor, but consumes less area. The linearity over the expected range of values to be stored in the capacitor is not heavily impacted by the nonlinearity. The storage capacitor also allows true correlated sampling (Figure 5.9).

5.4.3.2 Pixel Design

The data sheet of the process states that the dark current of a photodiode is 45 pA/cm². Thus, a 20 × 20 μm photodiode would have a dark current of 0.18 fA. The transistor sizes were chosen to balance the fill factor with speed. The column lines were assumed to have about 3 pF of resistance (an overestimation from process parameters and a function of column line width and length). Assuming that in the worst case, the

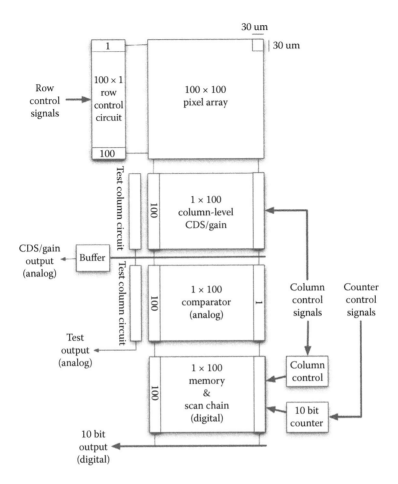

FIGURE 5.9   Diagram of a CMOS image sensor in Austria Microsystems' 0.35 μm OPTO.

column line will swing by 1 V and settle well within 1 μs, the NMOS follower was designed to source 6 μA of current. The same method was used to size the PMOS follower. All transistors have four times the minimum length of the process to minimize leakage (Figure 5.10).

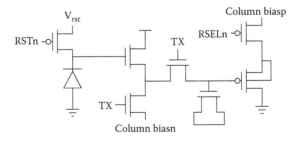

FIGURE 5.10   Pixel schematic.

The photodiode is reset through a PMOS, to ensure that there is no offset (Vth) on the reset voltage (Vrst). Two TX NMOS switches are implemented to reduce power consumption—the NMOS source follower is turned on only when the photodiode voltage (VPD) needs to be stored into the capacitor. The value stored on the capacitor is read through a PMOS source follower that is turned on only when the row is selected (RSELn = 0), saving power.

The normal operation of the pixel is as follows, assuming rolling shutter operation:

1. RSELn is set to 0 to output the capacitor value (VPD at reset) to the column CDS circuit.
2. TX is set to 1 to output the current photodiode voltage to the column CDS circuit.
3. CDS is performed and column-level ADCs convert the output values.
4. The ADC values are stored into column-level memory and read out to the PC.
5. RSELn is set to 1 to discontinue the pixel output to the column lines.
6. RSTn is pulsed 1 to 0 to 1 in order to reset the photodiode. This reset voltage is also transferred to the storage capacitor because TX is still 1.
7. TX is set to 0 to store the reset voltage for CDS operation after a given amount of time (integration time).
8. The shift scan register is set to the next row and the process is repeated.

5.4.3.3   Column Circuitry

Each column of the sensor has a CDS with three capacitors to enable different gain levels. A simple comparator is used to compare the CDS circuit output with an on-chip generated ramp and is essentially a single-slope ADC (Johns and Martin, 1996). Both the CDS and comparator circuits are based on a basic OTA design (Baker, 2010).

The memory element is a simple series of 10 D flip-flops (DFFs). Each DFF is connected to a corresponding bit value of a global 10-bit counter. When the ADC starts, the counter also starts. When the column comparator detects that the on-chip generated voltage ramp has become greater than the CDS output, a pulse is generated. This pulse is also connected to the DFFs, and the DFFs store the corresponding bit value of the counter at the time of the pulse. Thus, the ADC has a resolution of 10 bits.

Assuming a 500 fps operation with a rolling shutter, all column-level operations (CDS, ADC, write to memory, and readout) will have to be completed in 20 μs/row (Figure 5.11). This time frame can be broken down into the following:

1 μs: Read first value (reset voltage for CDS or previous value for temporal difference).

1 μs: Read second value (integrated voltage for CDS or temporal difference).

12 μs: ADC conversion and write to memory.

5 μs: Read out 100 columns (20 ns/column = 20 MHz clock).

1 μs: Wait time.

5.4.3.4   Global Circuitry

There are six global components on the sensor. The first is a scan shift register (SSR). It is a simple chain of 100 DFFs (equal to the number of row). The input to the first

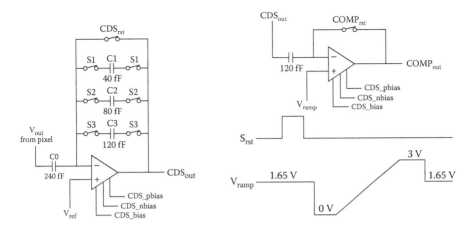

FIGURE 5.11   Column-level circuits and input timing for the single-slope ADC.

DFF is selectable by the user, but the subsequent DFFs have their inputs connected to the DFF before them. The output of each DFF corresponds to the row. If the first DFF outputs a 1, the first row is selected. If the second DFF outputs a 1, the second row is selected, and so on. The pixel control signals, generated by the on-chip state machine, are ANDed with the DFF outputs for each row.

Since the input of the first DFF can be selected, the SSR can be selected to offer a global or rolling shutter. For example, the clock to the SSR can be run for 100 cycles while the input is set to 1. This will result in all the DFFs outputting 1 (all rows being selected) at the end of the 100th clock cycle. Since the pixel control signals are ANDed with each row, the control signals will affect all rows. For rolling shutter operation, the input to the DFF is set to 1 for the first clock cycle. Thus, the 1 will travel through the SSR, 1 DFF/row at a time.

The second global component is a current mirror. An off-chip resistor, connected to a stable voltage source, provides the reference current to the current mirror. This current sets the drain and gate voltages of the NMOS transistor in the mirror, which sets the gate voltage of another NMOS transistor. The current generated is then mirrored to a PMOS network.

The current mirror then sources the current for the ramp generator. This current is then mirrored again (with gain) in the ramp generator. The mirrored current charges a fixed-value capacitor that was sized to reflect a charge time from 0 to 3.3 V in the range of 5–20 μs. Three slopes are possible by toggling different combinations of b0 and b1. When RSTCOMP is 1, the ramp generator outputs V rst, which is typically set to 1.656 V for optimal performance of the column-level comparator circuit (from simulation results). When RSTCOMP is 0 (ADC in operation), the output of the ramp generator is connected to the voltage of the capacitor, which is being charged linearly by the current mirrors. The output of the ramp generator is buffered by an OTA and is distributed to all column comparators of the sensor (Figure 5.12).

The bias generator also depends on the current mirror. The bias generator is based on a band gap voltage reference commonly shown in various CMOS textbooks

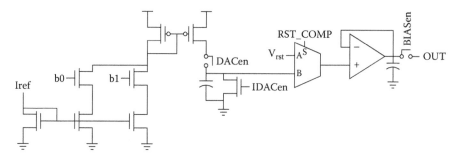

FIGURE 5.12 Ramp generator for the ADC.

(Baker, 1998; Johns and Martin, 1996). The transistors were sized to give appropriate voltages to the various components in the sensor.

The global synchronous counter is implemented with AND gates, XOR gates, and DFFs. The state machine is realized with two chains of DFFs. For each signal that needs to be generated on-chip, there is a chain of DFFs (basically, SSRs) that will push a pre-programmed logic level at each (state machine) clock cycle, which will be referred to as a state machine SSR. The state machine SSR is programmed by another SSR, referred to as programmable SSR. This was done to ensure that the chip would need to be programmed only once, since the state machine operation is destructive. After one run of the state machine, the contents of the state machine SSR will be empty. Thus, after each run of the state machine, the programmable SSRs (which store the desired pattern of the signal through each state) can transfer the values to the state machine SSR.

5.4.3.5 Sensor Characterization

First, the chip was tested with external biasing and a state machine, and using the on-chip ADC. The ideal performance was obtained with an 820,000 resistor and with the ramp generator's switches set to b0 = 1 and b1 = 0. This charges the ramp capacitor of 8 pF with 0.56 µA of current. Since the counter is running at 40 MHz, the capacitor charges from 0 to 1.78 V in 25.4 µs. With these settings, the maximum voltage swing per pixel is 1.24 V. The final photodiode has an area of 406 µm², which gives it a capacitance of 32.5 fF. Thus, the full well capacity is 253,000 e⁻, and the conversion gain is 4.9 µV/e⁻. In theory, the photodiode has a shot-noise-limited SNR of 54 dB.

The temporal noise components can be broken down into the following:

Photon shot noise: $\sqrt{253000e^-} = 503e^-$.
Reset noise: Negligible with CDS.
Pixel storage capacitor thermal noise (140 fF): 48 e⁻.
Dark shot noise (0.2 fA, 500 fps): 2.5 e⁻.
Column thermal noise (120 fF): 45 e⁻.

So the total noise can be given as

$$e_{rms} = \sqrt{(503e^-)^2 + (48e^-)^2 + (2.5e^-)^2 + (45e^-)^2 + (read\ noise)^2}$$

From experimental data, the total temporal noise is 523 e⁻ near saturation. Thus, the read noise is calculated to be 127 e⁻.

The quantum efficiency of the photodiode was measured using the same method as described in previous chapters. The exposed photodiode area is 393.1165 μm^2, giving the pixel a fill factor of 43.7%. Thus, the maximum operational QE (photodiode QE × fill factor) is 23%. The CDS gain was also tested. The analog output of the sensor was displayed and recorded with an oscilloscope under different settings. The gain settings of 2, 3, and 6× all worked well. However, from these experiments, it became apparent that the majority of the read noise is coming from the pixel. Since the gain is achieved at the column level, most of the read noise is also multiplied by the factor of the gain. Thus, while the column gain makes it easier to visualize the signal, it adds no benefit in suppressing noise.

5.4.3.6 Miniaturization

Due to the integration of many functionalities on-chip, the image sensor and auxiliary components fit onto a 10 × 10 mm PCB. The PCB contained bypass capacitors, a resistor, a voltage regulator, and a 24-pin Omnetics nanoconnector.

5.4.3.7 Experimental Results

Human embryonic kidney (HEK 293) cells were transiently transfected with an enhanced green fluorescent protein (EGFP)-based genetically encoded voltage sensor using Lipofectamine 2000 (Invitrogen, Grand Island, NY). Whole-cell patch clamp experiments were performed using the Patch Clamp PC-505B amplifier (Warner Instruments, Hamden, CT) with cells kept at the holding potential of –70 mV. Voltage-dependent changes in fluorescence intensity were provoked by 200 ms/+100 mV depolarizing steps recorded at 200 Hz. Change of fluorescence was imaged with a 60× 1.4 numerical aperture (N.A.) oil immersion objective lens using a 150 W xenon arc lamp (Opti-Quip, Highland Mills, New York). The image sensor was operated at 380 fps (Figure 5.13).

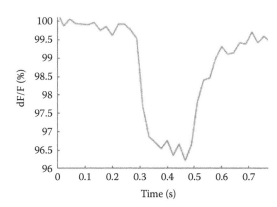

FIGURE 5.13 Change in fluorescence in a HEK 293 cell with an EGFP-based voltage sensor. Left: The HEK cell observed by the imaging system. Right: Change in fluorescence detected by the system. The curve is derived from a 2 × 2 pixel average of three trials.

5.4.3.8   Importance of Frame Rate and Limitation of Commercial Sensors

Due to the explosion of the CMOS image sensor market, many advanced noise reduction technologies, such as BSI, microlenses, and pinned photodiodes, have become common. Using commercial sensors (such as those from Omnivision and Aptina) can reduce design time. It is possible to achieve a performance in the range of 260 fps with 180×120 resolution and 46 dB SNR in a 10×11 mm PCB. However, while adequate for CSDI, better performance is required for VSDI.

Other miniature systems similar to the works presented in this chapter have used commercial sensors (Ghosh et al., 2011; Murari et al., 2010). However, these systems image at less than 100 fps. Even if they were able to detect small changes in fluorescence, the lack of speed would not give favorable results. An example is a HEK 293 cell in a similar experiment as above, but with a Super Ecliptic GFP-based voltage probe. In this particular example, +100 mV depolarizing steps were given with 5, 10, and 10 ms duration, at 100 ms intervals. Two sets of data were collected: one at 380 fps and another at 45 fps (the speed of Ghosh et al. (2011)). It is unclear whether the membrane potential of the cell changed, generating a change in fluorescence, at lower frame rates (Figure 5.14).

5.5   CONCLUSION AND POTENTIAL IMPROVEMENTS

The image sensors described in this chapter represent a compromise between small-cell phone-style cameras and large research-grade imaging systems. The integration of a bias generator, a state machine, and an ADC allows the sensor to be controlled using just three lines. Implementing a serial readout will require a faster read clock (400 Mhz), but will reduce data output lines from 10 to 1. Thus, the image sensor can be placed on a 10×10 mm PCB with an eight-wire connector. With the current design it is possible to include a full imaging and storage system that can be carried by a rodent.

While the sensor gives acceptable performance under high illumination, long-term studies using optical probes will require as little illumination as possible. In these situations, read noise becomes the major source of noise and experimental results have shown that the pixel read noise is the main component of read noise in the sensors.

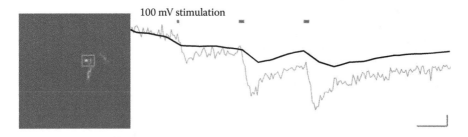

FIGURE 5.14   Left: The HEK cell observed by the imaging system. Right: Data captured at two different frame rates, single pixel with a five-trial average. Black curve is the data captured at 45 fps, and the gray curve is at 380 fps. The vertical bar represents a 2% change in fluorescence, and the horizontal bar represents 50 ms.

Due to the need to detect a change in intensity with 0.2% resolution, a SNR greater than 54 dB is required. Thus, a well capacity of at least 250,000 e$^-$ is required, since shot noise will be dominant near saturation. There are two approaches to further improve performance.

The first approach would be to improve upon a 3T structure (as done in the sensors described in this chapter). The advantage of the 3T pixel is its simplicity and large well size. However, for true CDS or global shutter operation, this structure requires a storage element that may take up a large area of the pixel, lowering the fill factor. However, microlensing can mitigate this fill factor issue, and bring the overall QE (QE × FF) close to maximum possible levels (QE of the PD). Another issue with 3T APS is the lower levels of QE compared with CCDs. With a lower QE, more photons are needed to reach saturation where the SNR is the highest. In order to minimize bleaching, a lower intensity of light is desired. Thus, QE needs to be as high as possible. Methods of increasing QE involve improving fabrication technologies and wafer postprocessing for backside illumination.

The second approach would be to use pinned photodiodes. These types of sensors can perform CDS without a large storage element in the pixel. Also, implementing microlensing and backside illumination has resulted in sensors with high sensitivity and low read noise. The drawback is that the well sizes are limited, due to the physical limitations of charge transfer in 4T structures. Neo sCMOS, for example, has a well capacity of 25,000 e$^-$, but the well capacity of most other pinned photodiode sensors is around 10,000 e$^-$. Thus, the SNRs achievable by these pixels (44 dB with sCMOS) are below the requirements for VSDI applications. However, binning the pixels can mitigate this problem.

Thus, BSI and microlensed pinned PD with binning and oversampling will give a better image due to its extremely low read noise, higher conversion gain, and higher QE. Such a system will also be better suited for optogenetic systems where the baseline signal level will be low (but changes in fluorescence larger with activity). Also, it is important to keep the pixel array as small as possible (at the cost of lower resolution), because a large area requires a larger working distance for the microscope optics, which in turn results in a taller/larger/heavier imaging system.

REFERENCES

Anaxagoras, T., Kent, P., Allinson, N., Turchetta, R., Pickering, T., Maneuski, D., Blue, A., and O'Shea, V. (2010). eLeNA: A parametric CMOS active-pixel sensor for the evaluation of reset noise reduction architectures. *IEEE Trans. Electron Devices*, 57(9):2163–2175.

Baker, R. J. (1998). References. In *CMOS circuit design, layout, and simulation*, 463–488. 2nd ed. New York: Wiley-IEEE Press.

Baker, R. J. (2010). Operational amplifiers I. In *CMOS circuit design, layout, and simulation*, 773–828. 3rd ed. New York: Wiley-IEEE Press.

Baker, B. J., Kosmidis, E. K., Vucinic, D., Falk, C. X., Cohen, L. B., Djurisic, M., and Zecevic, D. (2005). Imaging brain activity with voltage- and calcium-sensitive dyes. *Cell. Mol. Neurobiol.*, 25:245–282.

Berridge, M. J., Lipp, P., and Bootman, M. D. (2000). The versatility and universality of calcium signalling. *Nat. Rev. Mol. Cell Biol.*, 1:11–21.

Blasdel, G. G., and Salama, G. (1986). Voltage-sensitive dyes reveal a modular organization in monkey striate cortex. *Nature*, 321:579–585.

Bogaart, E., Hoekstra, W., Peters, I., Kleimann, A., and Bosiers, J. (2009). Very low dark current CCD image sensor. *IEEE Trans. Electron Devices*, 56(11):2462–2467.

Cohen, L., Keynes, R., and Hille, B. (1968). Light scattering and birefringence changes during nerve activity. *Nature*, 218:438–441.

Davis, D. J., Sachdev, R., and Pieribone, V. A. (2011). Effect of high velocity, large amplitude stimuli on the spread of depolarization in S1 "barrel" cortex. *Somatosens Mot Res.*

Ferezou, I., Bolea, S., and Petersen, C. C. (2006). Visualizing the cortical representation of whisker touch: Voltage-sensitive dye imaging in freely moving mice. *Neuron*, 50(4):617–629.

Ferezou, I., Haiss, F., Gentet, L. J., Aronoff, R., Weber, B., and Petersen, C. C. (2007). Spatiotemporal dynamics of cortical sensorimotor integration in behaving mice. *Neuron*, 56(5):907–923.

Flusberg, B. A., Nimmerjahn, A., Cocker, E. D., Mukamel, E. A., Barretto, R. P., Ko, T. H., Burns, L. D., Jung, J. C., and Schnitzer, M. J. (2008). High-speed, miniaturized fluorescence microscopy in freely moving mice. *Nat. Methods*, 5:935–938.

Fossum, E. R. (1993). Active pixel sensors: Are CCD's dinosaurs? In *Proceedings of SPIE: Electronic Imaging, Charge-Coupled Devices and Solid State Optical Sensors III*, vol. 1900, pp. 2–14.

Fujimori, I., and Sodini, C. (2002). Temporal noise in CMOS passive pixels. In *Proceedings of IEEE Sensors 2002*, vol. 1, pp. 140–145.

Ghosh, K. K., Burns, L. D., Cocker, E. D., Nimmerjahn, A., Ziv, Y., Gamal, A. E., and Schnitzer, M. J. (2011). Miniaturized integration of a fluorescence microscope. *Nat. Methods*, 8:871–878.

Helmchen, F., Fee, M. S., Tank, D. W., and Denk, W. (2001). A miniature head-mounted two-photon microscope: High-resolution brain imaging in freely moving animals. *Neuron*, 31(6):903–912.

Hewlett-Packard. (1998). Noise sources in CMOS image sensors. Technical report. Hewlett-Packard Components Group, Imaging Products Operation.

Ishihara, Y., and Tanigaki, K. (1983). A high photosensitive IL-CCD image sensor with monolithic resin lens array. In *1983 International Electron Devices Meeting*, vol. 29, pp. 497–500.

Janesick, J., Gunawan, F., Dosluoglu, T., Tower, J., and McCaffrey, N. (2002). Scientific CMOS pixels. *Exp. Astron.*, 14:33–43. 10.1023/A:1026128918608

Jerram, P., Burt, D., Guyatt, N., Hibon, V., Vaillant, J., and Henrion, Y. (2010). Back-thinned CMOS sensor optimization. In *Optical components and materials VII*, ed. S. Jiang, M. J. F. Digonnet, J. W. Glesener, and J. C. Dries, p. 759813. International Society for Optics and Photonics, 2010.

Johns, D., and Martin, K. (1996). *Analog integrated circuit design*. New Jersey: John Wiley & Sons, Inc.

Kasano, M., Inaba, Y., Mori, M., Kasuga, S., Murata, T., and Yamaguchi, T. (2006). A 2.0-µm pixel pitch MOS image sensor with 1.5 transistor/pixel and an amorphous SI color filter. *IEEE Trans. Electron Devices*, 53(4):611–617.

Kleinfelder, S., Lim, S., Liu, X., and El Gamal, A. (2001). A 10000 frames/s CMOS digital pixel sensor. *IEEE J. Solid-State Circuits*, 36(12):2049–2059.

Muller, R. S., Kamins, T. I., and Chan, M. (2002). *Device electronics for integrated circuits*. New York: John Wiley & Sons, Inc.

Murari, K., Etienne-Cummings, R., Cauwenberghs, G., and Thakor, N. (2010). An integrated imaging microscope for untethered cortical imaging in freely moving animals. In *2010 Annual International Conference of the IEEE Engineering in Medicine and Biology Society (EMBC)*, pp. 5795–5798.

Neher, E. (2000). Some quantitative aspects of calcium fluorimetry. In *Imaging neurons: A laboratory manual*. New York: Cold Spring Harbor Laboratory Press.

Osman, A., Park, J. H., Dickensheets, D., Platisa, J., Culurciello, E., and Pieribone, V. A. (2012). Design constraints for mobile, high-speed fluorescence brain imaging in awake animals. *IEEE Trans. Biomed. Circuits Syst.*, 6(5):446–453.

PerkinElmer. (2001). Photodiodes, phototransistors and infrared emitters.

Perry, R. (1999). Analysis and characterization of the spectral response of CMOS based integrated circuit (IC) photodetectors. In *Proceedings of the Thirteenth Biennial University/Government/Industry Microelectronics Symposium 1999*, pp. 170–175.

Petersen, C. C., Grinvald, A., and Sakmann, B. (2003). Spatiotemporal dynamics of sensory responses in layer 2/3 of rat barrel cortex measured in vivo by voltage-sensitive dye imaging combined with whole-cell voltage recordings and neuron reconstructions. *J. Neurosci.*, 23(4):1298–1309.

Rao, P. R., Munck, K. D., Minoglou, K., Vos, J. D., Sabuncuoglu, D., and Moor, P. D. (2011). Hybrid backside illuminated CMOS image sensors possessing low crosstalk. In *Sensors, systems, and next-generation satellites XV*, ed. R. Meynart, S. P. Neeck, and H. Shimoda, 81761D. Vol. 8176. SPIE.

Shoham, D., Glaser, D. E., Arieli, A., Kenet, T., Wijnbergen, C., Toledo, Y., Hildesheim, R., and Grinvald, A. (1999). Imaging cortical dynamics at high spatial and temporal resolution with novel blue voltage-sensitive dyes. *Neuron*, 24(4):791–802, ISSN 0896-6273, http://dx.doi.org/10.1016/S0896-627(00)81027-2

Siegel, M. S., and Isacoff, E. Y. (1997). A genetically encoded optical probe of membrane voltage. *Neuron*, 19:735–741.

Tasaki, I., Watanabe, A., Sandlin, R., and Carnay, L. (1968). Changes in fluorescence, turbidity, and birefringence associated with nerve excitation. *Proc. Natl. Acad. Sci. USA*, 61(3):883–888.

Tian, H., and El Gamal, A. (2001). Analysis of 1/f noise in switched MOSFET circuits. *IEEE Trans. Circuits Systems II*, 48(2):151–157.

Tian, H., Fowler, B., and Gamal, A. (2001a). Analysis of temporal noise in CMOS photodiode active pixel sensor. *IEEE J. Solid-State Circuits*, 36(1):92–101.

Tian, H., Liu, X., Lim, S., Kleinfelder, S., and Gamal, A. E. (2001b). Active pixel sensors fabricated in a standard 0.18μm CMOS technology. In *Proceedings of the SPIE*, 4306, Sensors and Camera Systems for Scientific, Industrial, and Digital Photography Applications II, 441–449. SPIE Press, doi 10.1117/12.426982.

Titus, A. H., Cheung, M. C.-K., and Chodavarapu, V. P. (2011). *Photodiodes—World activities in 2011*. InTech. Chapter 4, Photodiodes–World Activities in 2011, ISBN 978-953-307-530-3, In-Tech Publishing: Rijeka, Croatia.

Verhagen, J. V., Wesson, D. W., Netoff, T. I., White, J. A., and Wachowiak, M. (2007). Sniffing controls an adaptive filter of sensory input to the olfactory bulb. *Nat. Neurosci.*, 10:631–639. doi:10.1109/JSSC.1974.1050448.

Walden, R., Krambeck, R., Strain, R., McKenna, J., Scryer, N., and Smith, G. (1972). The buried channel charge coupled devices.

White, M., Lampe, D., Blaha, F., and Mack, I. (1974). Characterization of surface channel CCD image arrays at low light levels. *IEEE J. Solid-State Circuits*, 9(1):1–12. Nature Publishing Group, *Nature Neuroscience* 10, 631–639.

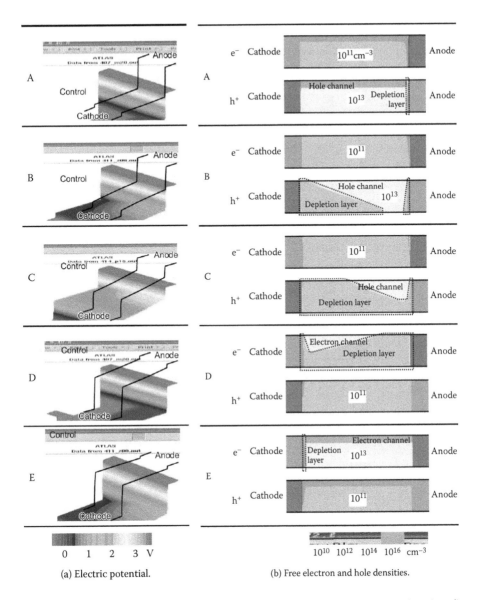

| | (a) Electric potential. | (b) Free electron and hole densities. |

0 1 2 3 V

10^{10} 10^{12} 10^{14} 10^{16} cm^{-3}

FIGURE 1.5

(continued)

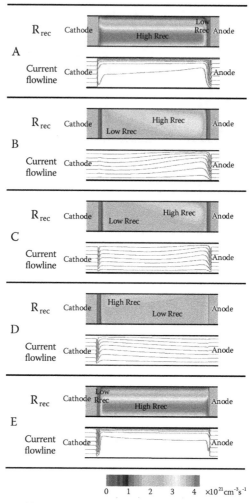

(c) Carrier recombination rate and current flowline

FIGURE 1.5 (CONTINUED)

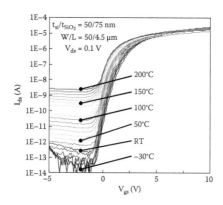

FIGURE 1.7 Temperature dependence of the transistor characteristic of the poly-Si TFT.

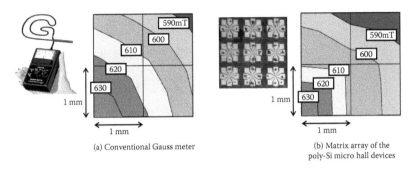

(a) Conventional Gauss meter

(b) Matrix array of the
poly-Si micro hall devices

FIGURE 1.17 Area sensing of magnetic field.

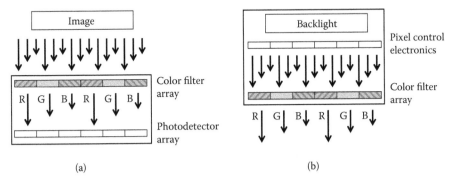

(a)

(b)

FIGURE 2.1 Simplified cross-sectional views of color image sensor (a) and color display devices (b). Arrays of thin-film color filters (red, green, blue), originally intended for sensory purposes, provide spectral sensitivities for analytical applications.

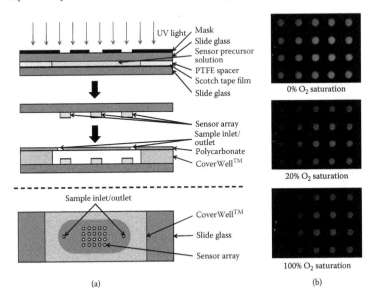

(a)

(b)

FIGURE 2.6 Optofluidic dissolved oxygen sensor assembly. (a) Photopatterning of PEG hydrogel array and layout of the assembled system. (b) Red-extracted images of PEG array in 0, 20, and 100% oxygen saturated water. (Reprinted from Park, J., W. Hong, and C.-S. Kim, *IEEE Sensors Journal*, 10(12), 1855–1861, 2010. Copyright © 2010 IEEE. With permission.)

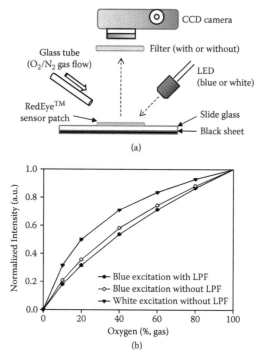

(a)

(b)

FIGURE 2.8 (a) The sensor imaging setup with a color CCD camera for gaseous oxygen quantification. A long-wave pass filter, blue LED, and white LED were selectively used for a series of measurements. The microscope setup is not shown. (b) Normalized Stern-Volmer plots to compare the performance of the three different imaging configurations. (Reprinted from Park, J., and C.-S. Kim, *Sensor Letters*, 9(1), 118–123, 2011. Copyright © 2011 American Scientific Publishers. With permission.)

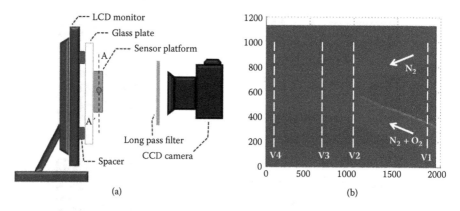

(a)   (b)

FIGURE 2.10 (a) Measurement setup with an LCD monitor as the excitation light source and a color camera as the photodetector. A color camera takes pictures of the fluidic sensor platform (8×8 cm^2) installed in close proximity to an LCD screen that provides an excitation blue light (470 nm) with uniform intensity over the sensor coating. (After Park, S., S. G. Achanta, and C.-S Kim, Fluorescence Intensity Measurements with Display Screen as Excitation Source, *Progress in Biomedical Optics and Imaging, Proceedings of SPIE*, Orlando, FL, April 25–29, 2011, Paper 802509. Permission required from SPIE.) (b) Stern-Volmer image of oxygen distribution (equivalent to I_0/I). *(continued)*

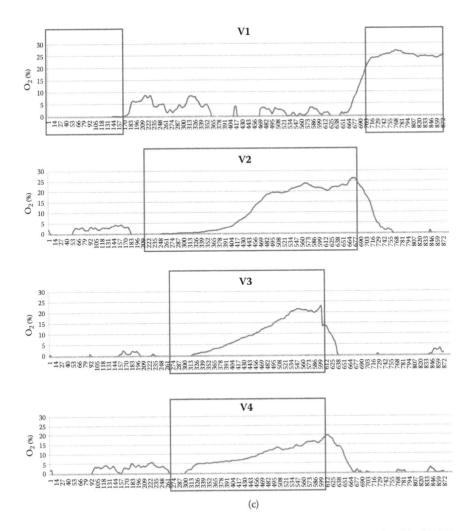

FIGURE 2.10 (CONTINUED) (c) Oxygen profiles at various locations defined in (b) (V1, V2, V3, and V4), showing a nitrogen and 20% oxygen fluxes at upper and lower branches, respectively.

FIGURE 3.4 An image divided into 28 regions using the median cut algorithm, with the centroids in each region represented as dots.

FIGURE 4.10 The cropped (320×240) original image (under zoom factor 48) with a target (red rectangle) is shown on the left. On the right, interested target points close to the human target in the segmented regions are selected and labeled (as L1–L9). The distance of the target to the camera is about 31 m.

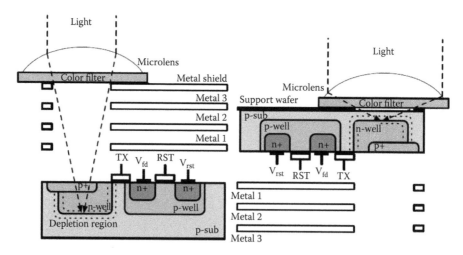

FIGURE 5.3 Comparison of different pixel designs: front-side illumination (left) and back-side illumination (right). For both, microlenses are shown. Transistors are not drawn to scale.

| (a) | (b) | (c) |

FIGURE 5.6 (a) Proposed implementation of the image sensor in a self-contained system for imaging fluorescent activity reporters in freely moving rodents. (b) The camera described in this chapter attached to a head-mountable microscope system; umbilical cable not shown. The microscope contains an excitation filter, emission filter, dichroic, and lenses (objective, condenser, relay). The light source contains a LED and heat sink. (c) Close-up of the camera housing and umbilical cable.

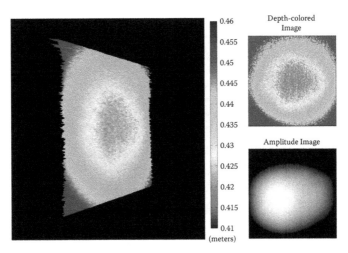

FIGURE 6.1 Example of an eye-in-hand configuration, with a ToF camera attached to the manipulator end effector, in this case a chlorophyll meter.

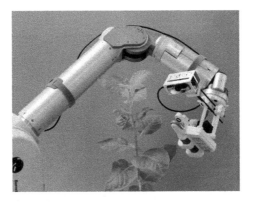

FIGURE 6.2 Typical raw ToF image of a flat surface at short distance. Depth, ranging from 0.41 to 0.46 m, is color coded. Observe that overillumination in the center leads to underestimation of depth (shift to red), while underillumination at borders causes overestimation (shift to blue).

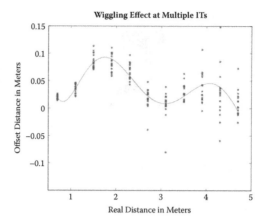

FIGURE 6.3 Depth distortion offset (wiggling effect). Blue: Measurements captured with a SR3100 ToF camera at several integration times (2–32 ms). Red: Six-degree polynomial approximated function.

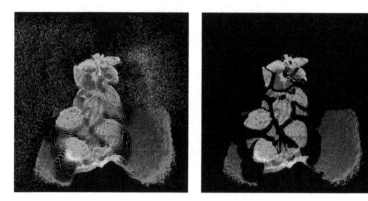

FIGURE 6.5 Reduction of noise by filtering pixels using a flying-point detector and depth threshold filtering.

FIGURE 6.6 Calibration errors produce bad colored points at the edge of the leaf. Additionally, observe the wrong color assignment in some background points, marked automatically in red, corresponding to the shadow of the leaf. This problem arises as the optical axes of the depth camera and the color camera are not the same and some depth points have no correspondence in the color image. Other sensor combinations, like Kinect, suffer the same problem.

(a)   (b)

FIGURE 6.7 Details of two different tools in the end effector: (a) a WAM robot and (b) a KUKA lightweight robot. Both tools require that the leaf is placed inside their lateral aperture. An eye-in-hand ToF camera permits acquiring the 3D plant structure required to compute robot motion.

FIGURE 6.8 Scene containing a detected leaf occlusion. After changing the point of view the occluded leaf is discovered and more characteristics (e.g., leaf area) can be measured. Depending on the particular leaf arrangement, it is not always possible to completely observe the occluded leaf.

FIGURE 6.10 Detail of a plant. Observe that the stems, even if they are thin, are correctly acquired.

FIGURE 6.11 Scene containing a possible merging of leaves. After changing the point of view, the ambiguity is clarified and two leaves are detected instead of one. Depending on the particular leaf arrangement, it is not always possible to completely disambiguate the occluded leaf.

FIGURE 7.40 The expected measurement results of the fabricated chip. Circles on the image indicate visual fixations, the size of the circle implies the fixation time, and arrows show the saccades.

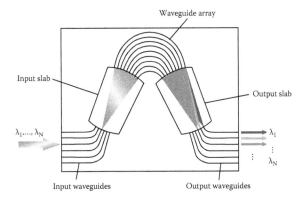

FIGURE 8.1 Basic optical circuit of AWG.

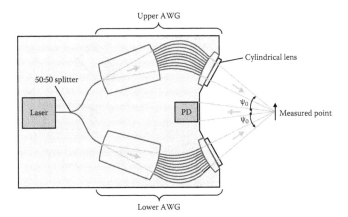

FIGURE 8.2 Optical circuit of wavelength-insensitive LDV using AWGs. (From K. Maru and Y. Fujii, *Appl. Mechanics Mater.*, 103, 76–81, 2011.)

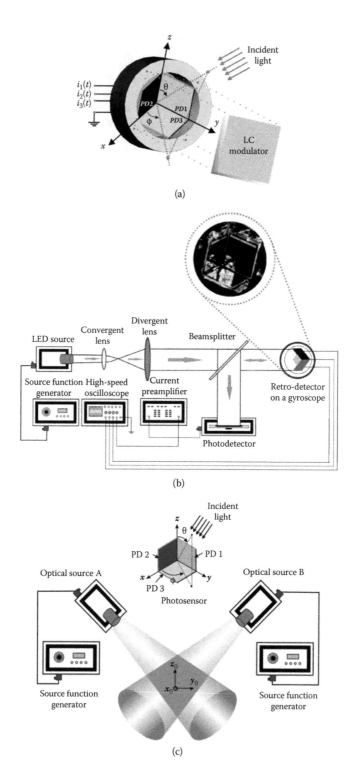

FIGURE 9.1 (a) Differential photosensor. (b) OWC system with a LED as the light source. The photosensor is shown in the inset. (c) OWL system with two optical sources.

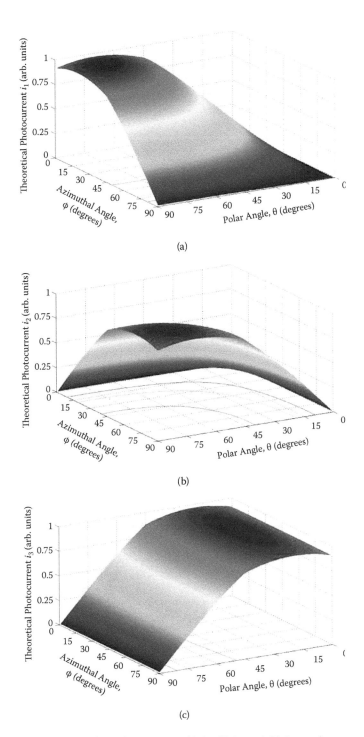

FIGURE 9.3 The theoretical photocurrents (a) i_1, (b) i_2, and (c) i_3 are shown as surfaces varying with φ and θ.   *(continued)*

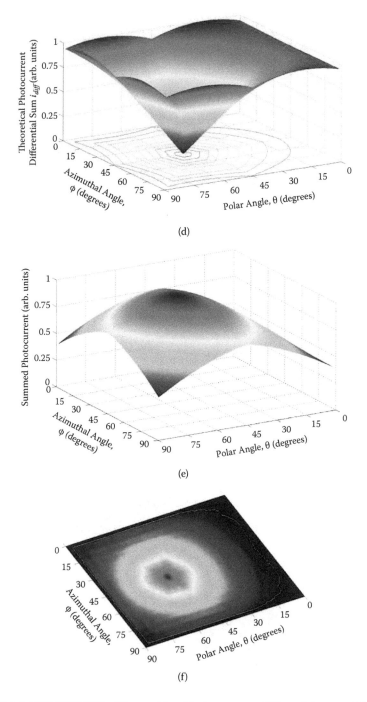

FIGURE 9.3 (CONTINUED) The resulting (d) photocurrent differential sum and (e) absolute summed photocurrent are also shown. (f) Two-dimensional view of the retroreflected power.

6 ToF Cameras for Eye-in-Hand Robotics*

G. Alenyà, S. Foix, and C. Torras

CONTENTS

6.1 INTRODUCTION

A time-of-flight (ToF) camera is a relatively new type of sensor that delivers three-dimensional images at high frame rate, simultaneously providing intensity data and

---

* Some parts of this chapter were originally published in G. Alenyà, S. Foix, and C. Torras, ToF Cameras for Active Vision in Robotics, *Sensors and Actuators A: Physical*, 218, 10–22, 2014.

range information for every pixel. It has been used in a wide range of applications, and here we will describe the lessons learned by using such cameras to perform robotic tasks, specifically in eye-in-hand configurations. In such configurations, the camera is attached to the end effector of a robot manipulator, so that new images can be obtained by actively changing the point of view of the camera (Figure 6.1).

In an eye-in-hand scenario, some particular characteristics of the sensor system are appreciated—mainly, the compactness and the detection in a short range, besides the obvious requirement of quality (precision and accuracy) in the obtained data. On the one hand, operation in a short range is desired because manipulator robots typically have a limited workspace, and the distance from the end effector to an object located in front of the robot is short. As will be demonstrated later, ToF cameras exhibit good performance in short ranges. On the other hand, as the sensor system is mounted on a robot arm it has to be lightweight, with no mobile parts, and as small as possible to avoid interference with the environment or the robot itself. ToF cameras fit this description well, as they are usually lightweight, have no mobile parts, and can be compact and small. Section 6.2 introduces ToF cameras and presents a critical comparison with RGBD cameras (Kinect), a different 3D sensor that is more and more commonly used in robotics.

Eye-in-hand configurations have been used extensively for object modeling [1], and more recently to enable robot interaction with the environment [2]. Section 6.3 presents and places in context some of the relevant works.

Regarding the quality of data, it is well known that raw ToF data are quite noisy and prone to several types of disturbances [3]. Some of them are systematic and can be calibrated, and others are nonsystematic and sometimes can be filtered out. In Section 6.4 systematic and nonsystematic error sources are reviewed. Section 6.5 shows the combination of both ToF and color images to obtain colored point clouds.

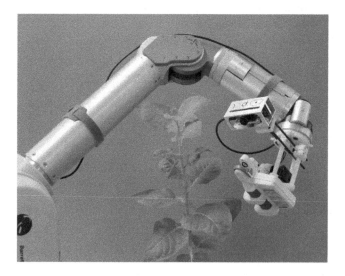

FIGURE 6.1   (See color insert.) Example of an eye-in-hand configuration, with a ToF camera attached to the manipulator end effector, in this case a chlorophyll meter.

The ability to actively move the camera depending on the scene provides some advantages. In Section 6.6 we show three illustrative examples: understanding the 3D structure of some relevant parts of the scene to enable robot-object interaction, obtaining detailed views of 3D structures, and disambiguation to enhance segmentation algorithms. Finally, some conclusions are drawn in Section 6.7.

6.2   ToF CAMERAS

In ToF cameras depth measurements are based on the well-known time-of-flight principle. A radio-frequency-modulated light field is emitted and then reflected back to the sensor, which allows for the parallel measurement of its phase (cross-correlation), offset, and amplitude [4]. Figure 6.2 shows a typical raw image of a flat surface with the depth values coded as different color values.

The main characteristics of two ToF sensors, PMD CamCube 3 and Mesa Swissranger 4K, are detailed in Table 6.1. We also include the specifications of the Kinect sensors to compare with a very common alternative 3D sensor. Both camera types can deliver depth images at reasonably high frame rates. The main difference is in resolution: ToF cameras still have limited resolution (typically around 200×200), while the Kinect depth camera exhibits VGA resolution. Both camera types are autoilluminated, so in principle they can work in a wide variety of illumination conditions.

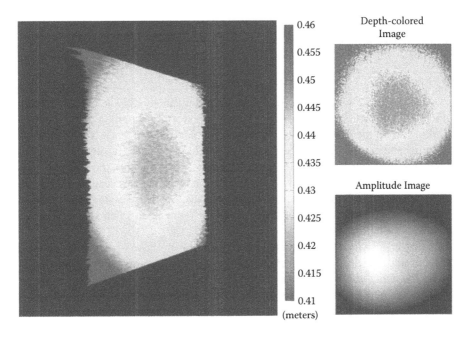

FIGURE 6.2   (See color insert.) Typical raw ToF image of a flat surface at short distance. Depth, ranging from 0.41 to 0.46 m, is color coded. Observe that overillumination in the center leads to underestimation of depth (shift to red), while underillumination at borders causes overestimation (shift to blue).

TABLE 6.1

Specifications of Different ToF Cameras, and Comparison with Kinect Features

Camera Model	PMD CamCube	Swissranger 4K	Kinect
Technology	ToF	ToF	Structured light
Image size	200×200	176×144	640×480 (depth)
			1280×1024 (color)
Frame rate	40 fps	30 fps	30 fps (depth)
	Up to 80 fps	Up to 50 fps	30/15 fps (color)
Lens	CS mount $f = 12.8$	Standard/wide option	Fixed
Range	0.3–7 m	0.8–5 m	0.5–3.5 m
		0.8–8 m	
Field of view	40×40	43.6×34.6	57×43
		69×56	
Focus	Adjustable	Adjustable	Fixed
Integration time	Manual	Manual	Auto
Illumination	Auto	Auto	Auto (depth)
Outdoor	Suppression Background Illumination	No	No
Images	Depth	Depth	Depth
	Intensity	Intensity	Color
	Amplitude	Amplitude	
	Confidence	Confidence	
Interface	USB	USB—Ethernet	USB

We focus this review on 3D perception for robotic manipulation and object modeling; thus, resolution is an important factor. It is worth mentioning that the closest working depth for Kinect is 0.5 m,[1] whereas that for ToF can reach 0.3 m, and even 0.2 m when equipped with new illumination units [5]. Kinect resolution is higher, but closer views can be obtained with ToF cameras. Consequently, the resulting horizontal (or vertical) resolution in mm per pixel of both cameras is very similar, as the lower resolution of ToF cameras can be compensated with closer image acquisition. The major consequence is that the density of the point cloud when viewing a given object is similar for both camera types.

However, placing ToF cameras closer to the object has two problems, related to focus and integration time. Like any other camera that uses optics, focus determines the depth of field (distance range where sharp images are obtained). If we set the focus to obtain sharp images of closer objects, then the depth of field is small. ToF cameras do not have autofocus capabilities, so the focus (and consequently the desired depth of field) has to be determined in advance.

Moreover, integration time has to be manually adjusted. Integration time has a strong impact on the quality of the obtained images, and each integration time sets the camera for a specific range of depths. As before, for close distances the range of possible depths for a given integration time is small.

Some of the ToF cameras have the capability of autoadjusting the integration time. However, depth calibration of ToF cameras is dependent on the current integration time, and a common practice is to calibrate for only one integration time, which is manually determined depending on the expected depth range.

One of the advantages of Kinect is the ability of delivering colored depth points if required. As will be presented in Section 6.5, coloring ToF depth points is also possible but requires some additional efforts.

One common problem with both cameras is that they do not provide a dense depth map. The delivered depth images contain holes corresponding to the zones where the sensors have problems, whether due to the material of the objects (reflection, transparency, light absorption) or their position (out of range, with occlusions). Kinect is more sensitive to this problem by construction.

Finally, we have tested ToF cameras in outdoor scenarios with sunlight [6]. An algorithm has been proposed to select the best integration time depending on the sun conditions, as well as a suitable strategy to combine two frames to obtain depth images even when a plant is partially illuminated with direct sunlight and partially in shadow, as is common in greenhouses. As could be expected, a ToF camera provides depth information, but with more noisy depth readings in parts exposed to direct sunlight.

6.3   USING ToF CAMERAS IN ROBOTIC MANIPULATION TASKS

ToF cameras have been used to sense relatively large depth values for mapping or obstacle avoidance in mobile robotics, and also for human detection and interaction. At closer distances, ToF cameras have been applied to object modeling [7, 8], precise surface reconstruction [9], and to grasp known [10] and unknown [11] objects. We focus our review on two complementary areas: scene-related tasks and object-related tasks. *Scene-related tasks* generally involve moving the camera using a mobile robot. Although the range of distances involved is rather long, the techniques and ideas can be applied to eye-in-hand algorithms. *Object-related tasks* involve the use of ToF cameras at close distances. The most common application is object modeling, and to a lesser extent to enable object manipulation.

A table is provided in each section to summarize and give a comprehensive view of i0074s contents. Our conclusion is that the most exploited feature of ToF cameras is their capability to deliver complete scene depth maps at a high frame rate without the need for moving parts. Moreover, foreground/background segmentation methods based on depth information are quite straightforward, so ToF images are used in many applications requiring them. A good characteristic is that geometric invariants as well as metric constraints can be naturally applied to ToF depth images.

The depth-intensity image pair is also often used, exploiting the fact that both images are delivered already registered. In applications where the reduced resolution of a ToF camera is critical, it is complemented with other sensors, usually color cameras. ToF cameras are used in human environments because they are eye-safe and permit avoiding physical contact and dedicated markers or hardware.

Some of the reviewed works do not apply any calibration method to rectify the depth images. We believe that this explains several of the errors and inaccuracies reported in some experiments, and that with proper calibration better results can be obtained. We note that ToF technology is evolving and depth correction methods are still subject to investigation.

The first works that appeared were comparisons between ToF and other technologies. Then, in subsequent works, these technologies were gradually complemented, and sometimes substituted, by ToF sensors.

6.3.1 SCENE-RELATED TASKS

This kind of application deals with tasks involving scenes that contain objects like furniture and walls. Observe that the expected range of distances to these objects is relatively wide. A usual framework in these applications is to install the camera on a mobile robot and use it for robot navigation and mapping. As will be seen, one of the areas where ToF sensors are adequate is in obstacle avoidance, because the detection region is not only horizontal (like in laser scanners) but also vertical, allowing the robot to detect obstacles with complex shapes. Clearly, *the most appreciated characteristic of ToF sensors here is the high frame rate* (see Table 6.2). Some applications also benefit from the metric information obtained with depth images.

6.3.1.1 Comparison

Initial works were devoted to the comparison of ToF with other sensors, mainly laser scanners. Thanks to the larger vertical field of view of ToF cameras, difficult obstacles (like tables) are better detected by them than by 2D laser scanners. For example, Weingarten et al. [12] demonstrated this in the context of an obstacle avoidance algorithm.

To obtain a comparable detection area, a 3D scanner can be built from a pivoted 2D laser scanner. May et al. [13, 14] compared the performance of their robot navigation algorithm using such a sensor and a ToF camera. One of the main difficulties they encountered is the accumulated error in the map created with the ToF camera, leading to failures when closing loops, for instance. Compared to pivoted laser scanners, accumulated errors usually occur more often with ToF cameras due to their smaller field of view. As we will see in the next section, this problem is also present in object modeling tasks.

6.3.1.2 Only ToF

ToF sensors have been used successfully as the unique sensor in some mobile robotic applications, despite their characteristic limited resolution. For mapping purposes, *ToF sensors are very interesting because they allow extraction of geometric features.* Most of the reviewed applications extract planar regions using both intensity and depth images. In [15], May et al. explored different methods to improve pose estimation. They additionally proposed a final refinement step that involves the alignment of corresponding surface normals leading to improved 3D scene maps computed at frame rate. The normal of the extracted planes was also used by Hedge and Ye [16] to detect badly conditioned plane detection, like horizontal planes in a staircase. Also, Pathak et al. [35] reported the use of ToF to extract planes for 3D mapping.

TABLE 6.2

ToF Camera Usage in Scene-Related Tasks

Reference	Topic	Advantages	Type of Sensor
Weingarten et al. [12]	Obstacle avoidance in static environment	3D at high rate	SR2 (depth)
May et al. [13, 14]	3D mapping	3D at high rate/no required pan-tilt	SR2 (depth)
May et al. [15]	Pose estimation/3D mapping	Registered depth-intensity	SR3 (depth + intensity)
Hedge and Ye [16]	Planar feature 3D mapping	3D at high rate/no required pan-tilt	SR3
Ohno et al. [17]	3D mapping	3D at high rate	SR2
Stipes et al. [18]	3D mapping/point selection	Registered depth-intensity	SR3
May et al. [19]	3D mapping/SLAM	3D at high rate	SR3
Gemeiner et al. [20]	Corner filtering	Registered depth-intensity	SR3 (depth + intensity)
Thielemann et al. [21]	Navigation in pipelines	3D allows geometric primitives search	SR3
Sheh et al. [22]	Navigation in hard environment	3D at high rate	SR3 + inertial
Swadzba et al. [23]	3D mapping in dynamic environment	3D at high rate/registered depth-intensity	SR3 (depth + intensity)
Acharya et al. [24] Gallo et al. [25]	Safe car parking	Improved depth range/3D at high rate	Canesta
Goktuk and A. Rafii [26]	Object classification (air bag application)	Light/texture/shadow independence	Canesta
Yuan et al. [27]	Navigation and obstacle avoidance	Increased detection zone	SR3 + laser
Kuhnert and Stommel [28]	3D reconstruction	Easy color registration	PMD + stereo
Netramai et al. [29]	Motion estimation	3D at high rate	PMD + stereo
Huhle et al. [30]	3D mapping	Easy registration of depth and color	PMD + color camera
Prusak et al. [31]	Obstacle avoidance/ map building	Absolute scale/better pose estimation	PMD + spherical camera
Swadzba et al. [32]	3D mapping/map optimization	3D at high rate	SR3
Vaskevicius et al. [33]	Localization/map optimization	Neighborhood relation of pixels	SR3
Poppinga [34]		No color restrictions	

Alternatively, the acquired crude point clouds can be processed by a variant of the iterative closest point (ICP) algorithm to find the relation between two point clouds. For example, a real-time 3D map construction algorithm is proposed by Ohno et al. [17] in the context of a snake-like rescue robot operating in complex environments, like rubble in disaster-like scenarios. Here, a modification of the classical ICP algorithm is proposed to cope with ToF noisy readings and to speed up the process.

Another adaptation of an ICP-like algorithm for ToF images was presented by Stipes et al. [18], where both the depth and intensity images were used. They presented a probabilistic point sampling process to obtain significant points used in the registration process.

ICP assumes that both point clouds overlap, so wrong depth points can distort the result. May et al. [19] presented an ICP variant to take this explicitly into account. They proposed a mapping algorithm using a simultaneous localization and mapping (SLAM) technique to reduce the reconstruction error that is especially useful when a zone of the scenario is revisited, i.e., when closing a loop.

Also, with potential applications to SLAM, Gemeiner et al. [20] proposed a corner-filtering scheme combining both the intensity and depth images of a ToF camera.

Complex environments are a good test field for ToF sensors, as they are capable of naturally recovering their geometry. In the context of pipeline inspection, Thielemann et al. [21] proposed using a ToF camera to detect the different junctions based not on appearance but on geometric properties. Here the self-illumination mechanism of ToF sensors is appreciated. Furthermore, Sheh et al. [22] proposed a ToF-based navigation system for a random stepfield terrain.[2] They used the depth information to color an array of pixels and then performed some classical edge detection algorithms in this array, which is called *heightfield*. The heading and attitude compensation of the image is performed using an inertial unit.

ToF sensors have proved to also be applicable in dynamic environment mapping thanks to their characteristic high frame rate. Swadzba et al. [23] presented a scene reconstruction algorithm that discards dynamic objects, like pedestrians, using a static camera in the difficult case of short sequences (2–3 s). Motion is recovered via optical flow in the intensity images, and then transferred to the depth image to compute a 3D velocity vector.

ToF cameras have also been employed in the automotive field to assist in parking operations. In [24] Acharya et al. described the system design of a ToF camera for backup obstacle detection. In [25] the same group presented an application of a similar camera for the detection of curves and ramps in parking settings. A modified Ransac algorithm, which uses only the best inliers, is used to find the best fitting of the planar patches that model the environment. ToF has also been used to control the deployment of the air bag system, depending on the nature of the occupant in a car [26]: adult, child, child seat, or objects.

6.3.1.3   Fusion with Other Sensors

Some other authors have recently started to fuse ToF cameras with other sensors, i.e., laser scanners and different types of color cameras. A simple approach is to integrate ToF into existing algorithms. For example, Yuan et al. [27] proposed a fusion process to integrate 3D data in the domain of laser data by projecting ToF point clouds onto the laser plane. This is applicable when considering a simple-shaped robot, i.e., one that can be approximated by a cylinder, and it entails a minimum update of their previous laser-scanner-based algorithm. Nevertheless, the resulting algorithm can cope with new kinds of obstacles in a simple way. Note that this is not a pure 3D approach, and it is not using the potentiality of having full 3D information at a high frame rate.

Fusion of color and depth information in scene tasks seems to have great potential. In a preliminary work, Kuhnert and Stommel [28] presented a revision of their 3D environment reconstruction algorithm combining information from a stereo system and a ToF sensor. Later, Netramai et al. [29] compared the performance of a motion estimation algorithm using both ToF and depth from stereo. They also presented an oversimplified fusion algorithm that relies on the optical calibration of both sensors to solve the correspondence problem. These works propose fusion paradigms combining the results produced in two almost independent processes.

Contrarily, Huhle et al. [30] presented a color ICP algorithm useful for scene-based image registration, showing that introducing color information from a classical camera in the beginning of the process effectively increases the registration quality.

Depth information allows us to identify in a robust manner not only obstacles but also holes and depressions. Prusak et al. [31] proposed a joint approach to pose estimation, map building, robot navigation, and collision avoidance. The authors used a PMD camera combined with a high-resolution spherical camera in order to exploit both the wide field of view of the latter for feature tracking and pose estimation, and the absolute scale of the former. The authors relied on a previous work on integration of 2D and 3D sensors [36, 37], showing how restrictions of standard structure-from-motion approaches (mainly scale ambiguity and the need for lateral movement) could be overcome by using a 3D range camera. The approach produced 3D maps in real time, up to 3 fps, with an ICP-like algorithm and an incremental mapping approach.

6.3.1.4 Noisy Data Enhancement

Swadzba et al. [32] proposed a new algorithm to cluster redundant points using a virtual plane, which apparently performs better in planar regions and reduces noise, improving registration results. Furthermore, a group at Jacobs University [33, 34] proposed to identify surfaces using a region-growing approach that allows the poligonization of the resulting regions in an incremental manner. The nature of the information delivered by ToF cameras, specially the neighborhood relation of the different points, is explicitly exploited, and also their noisy nature is taken into account. Moreover, some comparisons with results from stereo rigs are reported.

Finally, Huhle et al. [38] proposed an alternative representation of the map by means of the normal distribution transform, which efficiently compresses the scan data, reducing memory requirements. This representation seems to also be well suited for the typical noisy ToF depth images.

6.3.2 OBJECT-RELATED TASKS

ToF cameras have also been successfully used for object and small surface reconstruction, where the range of distances is small. A comprehensive summary is given in Table 6.3.

TABLE 6.3

ToF Camera Usage in Object-Related Tasks

Reference	Topic	Advantages	Type of Sensor
Ghobadi et al. [39]	Dynamic object detection and classification	Color and light independence	PMD
Hussmann and Liepert [40]	Object pose	Easy object/background segmentation	PMD
Guomundsson et al. [41]	Known object pose estimation	Light independent/ absolute scale	SR3
Beder et al. [42]	Surface reconstruction using patchlets	ToF easily combines with stereo	PMD
Fuchs and May [9]	Precise surface reconstruction	3D at high rate	SR3/O3D100 (depth)
Dellen et al. [7] Foix et al. [8]	3D object reconstruction	3D at high rate	SR3 (depth)
Kuehnle et al. [10]	Object recognition for grasping	3D allows geometric primitives search	SR3
Grundmann et al. [43]	Collision-free object manipulation	3D at high rate	SR3 + stereo
Reiser and Kubacki [44]	Position-based visual servoing	3D is simply obtained/ no model needed	SR3 (depth)
Gächter et al. [45]	Object part detection for classification	3D at high rate	SR3
Shin et al. [46]			SR2
Marton et al. [47]	Object categorization	ToF easily combines with stereo	SR4 + color
Saxena et al. [11]	Grasping unknown objects	3D at high rate	SR3 + stereo
Zhu et al. [48]	Short-range depth maps	ToF easily combines with stereo	SR3 + stereo
Lindner et al. [49]	Object segmentation for recognition	Easy color registration	PMD + color camera
Fischer et al. [50]	Occlusion handling in virtual objects	3D at high rate	PMD + color camera

6.3.2.1 Comparison with Stereovision

A classical solution in the area of object modeling is the use of calibrated stereo rigs. Therefore, initial works were devoted to their comparison with ToF sensors showing the potential of the latter when poorly textured objects are considered, and when background-foreground segmentation is difficult. For planar and untextured object surfaces, where stereo techniques clearly fail, Ghobadi et al. [39] compared the results of a dynamic object detection algorithm based on support vector machine (SVM) using stereo and ToF depth images. In the same manner, Hussmann and Liepert [40] compared ToF and stereovision for object pose computation. The key difference favorable to the ToF camera is its ability to effectively segment the object and the background, even if their color or texture is exactly the same (i.e., a white object on a white table). They also proposed a simple method to obtain object pose from a depth image.

Another comparison was presented by Guomundsson et al. [41]. They classified and estimated the pose of some simple geometric objects using a local linear embedding (LLE) algorithm, and contrasted the results of using the intensity image and the depth image. Their analysis shows that range data add robustness to the model and simplifies some preprocessing steps, and in general, the generated models better capture the nature of the object. Stereo and ToF were also compared by Beder et al. [42] in the framework of surface patchlet identification and pose estimation. In their setup, using a highly textured surface for stereo experiments, ToF slightly outperformed stereo in terms of depth and normal direction to the patchlet. Thus, ToF can be used to benchmark stereo surface reconstruction algorithms.

6.3.2.2 ToF for Surface Reconstruction

To obtain 3D object surfaces, multiple 3D images need to be acquired and the resulting 3D point clouds should be combined. The setups for these object modeling algorithms usually include a ToF camera mounted on the end effector of a robotic arm. *Point cloud registration is more critical in object modeling than in scene modeling.* Even if the hand-eye system is precisely calibrated, the displacement given by the robot is usually not enough and the transformation between different point clouds has to be calculated. The application of ICP in two consecutive views naturally accumulates errors, and consequently more precise algorithms need to be used.

To obtain precise object models, Fuchs and May [9] performed a circular trajectory around the object to acquire equally spaced images, and used a *simultaneous matching* algorithm [51] instead of classical ICP to distribute the errors in all the estimated displacements. Their work also included a comparison of two different ToF cameras. Alternatively, Dellen et al. [7] proposed a fine registration algorithm based on an ICP algorithm using invariant geometric features. The resulting model is obtained after reducing noise and outliers by treating the coarse registered point cloud as a system of interacting masses connected via elastic forces. Alternatively, Foix et al. [8] proposed a method to compute the covariance of the point cloud's registration process (ICP), and applied an iterative view-based aggregation method to build object models under noisy conditions. Their method does not need accurate hand-eye calibration since it uses globally consistent probabilistic data fusion by means of a view-based information form SLAM algorithm, and can be executed in real time, taking full advantage of the high frame rate of the ToF camera.

6.3.2.3 ToF for Object Manipulation

Object recognition and object pose estimation algorithms are usually related to robotic manipulation applications: objects have to be identified or categorized with the aim of finding and extracting some characteristics to interact with them. This is usually a challenging task, as ToF depth images are noisy, and low sensor resolution leads to only a few depth points per object.

Kuehnle et al. [10] explored the use of a ToF camera to recognize and locate 3D objects in the framework of the robotic manipulation system DESIRE. Objects were modeled with geometric primitives. Although they used depth images rectified up to some level, their system was not reliable enough. In a subsequent work [43] they used the ToF camera to detect unknown objects and classify them as obstacles, and used a stereo camera system to identify known objects using SIFT features. As is widely known, this second approach required textured objects, while their first approach did not. In the same project, Reiser and Kubacki [44] proposed a method to actively orientate the camera using a visual servoing approach to control a pan-and-tilt unit. They proved that position-based visual servoing is straightforward by using a ToF camera, because of its ability to deliver 3D images at high rates.

In a different way, Gächter et al. [45] proposed to detect and classify objects by identifying their different parts. For example, chairs are modeled by finding their legs, which in turn are modeled with vertical bounding boxes. The tracking of the different parts in the image sequence is performed using an extended particle filter, and the recognition algorithm is based on a SVM, which proves again to be useful in typical noisy ToF images. Later, Shin et al. [46] used this incremental part detector to propose a classification algorithm based on a geometric grammar. However, they used a simulated environment because the classification in real scenarios does not seem to be reliable enough.

Depth information is very useful in cluttered environments to detect and grasp unknown objects: the 3D region of interest can be extracted easily, and some object segmentation algorithms can be developed combining cues from both a ToF sensor and a color camera. Using such a combined sensor, Marton et al. [47] proposed a probabilistic categorization algorithm for kitchen objects. This work used a new SR4000 camera. The sensor assigns a confidence value to each depth reading, allowing us to infer if the object material is producing bad sensor readings.

Thanks to the depth information, some grasping properties can be easier to evaluate, i.e., form- and force-closure, sufficient contact with the object, distance to obstacles, and distance between the center of the object and the contact point. Saxena et al. [11] used this advantage to propose a learning grasp strategy that identifies good grasping points using partial shape information of unknown objects. The contribution of the depth information allows us to update an already presented method using a color camera, with the advantage of having depths even in textureless portions of the objects.

6.3.2.4 Fusion Algorithms

In fact, ToF and stereo systems naturally complement one another. As has been argued before, ToF performs correctly in poorly textured surfaces and object segmentation becomes easy even in poorly contrasted situations. Contrarily, it has difficulties precisely in textured surfaces and short distances, where stereo outperforms it. This fact has been exploited in several works. For example, Zhu et al. [48] proposed a probabilistic framework to fuse depth maps from stereo and the ToF sensor. They used a depth calibration method to improve the ToF image, which is useful in small depth ranges (from 1 to 1.4 m).

Another fusion framework was proposed by Lindner et al. [49] using calibration and scaling algorithms. They obtained a dense colored depth map using the geometrical points' correspondence between the ToF and color cameras by assigning a color to the ToF depth points, and interpolating the depth of the rest of the color camera pixels. A way to detect areas not seen by the color camera is also provided, as well as some techniques to enhance edges and detect invalid pixels.

Finally, in the context of augmented reality, Fischer et al. [50] combined a ToF camera and a standard color camera to handle virtual object occlusions caused by real objects in the scene. Fast 3D information is highly valuable, as well as its independence on lightning conditions, object texture, and color. They did not use any depth calibration or noise outlier removal algorithm, and consequently, the negative effect of noise is clearly visible in their results.

6.3.2.5 Summary and Final Remarks

ToF cameras have been successfully used for object and small surface reconstruction at close distances. In general, the scenario for these applications involves a robotic manipulator and the task requires modeling object shape. In such settings, one has to expect that some oversaturation problems may occur when acquiring depth images. On the contrary, as the range of depths is short, calibration can be simplified.

Some of the reviewed works do not apply any calibration method to rectify the depth images. We believe that this explains some of the errors and inaccuracies reported in some experiments, and that with proper calibration, better results can be obtained. We note that ToF technology is evolving and depth correction methods are still subject to investigation.

Foreground/background segmentation methods based on depth information are quite straightforward, so ToF images are used in many applications requiring them. A good characteristic is that geometric invariants as well as metric constraints can be naturally used with the ToF depth images.

ICP-like techniques are the preferred solution to reconstruct surfaces. A common approach to identify objects is the use of support vector machines, which perform adequately when considering the noisy point models obtained with one ToF image or when merging different ToF views.

The high frame rate of ToF sensors is a key advantage, but also the natural combination with color cameras and stereo rigs. The fact that the depth and intensity images are delivered already registered is handy in some contexts, but in applications where the reduced resolution of a ToF camera is critical, it is complemented with other sensors, usually color cameras. Actually, a growing trend is observed not to use the intensity image supplied by the ToF camera, preferring the combination with high-resolution conventional cameras.

6.4 DEPTH MEASUREMENT ERRORS

Raw measurements captured by ToF cameras provide noisy depth data. Default factory calibration can be used in some applications where accuracy is not a strong requirement and the allowed depth range is very large. For the rest of the applications ToF cameras have to be specifically calibrated over the defined application depth range. Two types of errors, *systematic* and *nonsystematic*, can interfere and consequently corrupt ToF depth readings. Two of the most important systematic errors are *depth distortion*, an offset that affects all images and is dependent on the measured depth (Figure 6.3), and *built-in pixel errors*, which are a constant offset of each pixel independent of the measured depth. While systematic errors are compensated by calibration, nonsystematic ones are minimized by filtering.

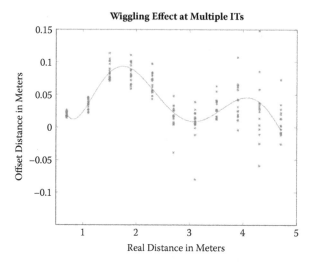

FIGURE 6.3 (See color insert.) Depth distortion offset (wiggling effect). Blue: Measurements captured with a SR3100 ToF camera at several integration times (2–32 ms). Red: Six-degree polynomial approximated function.

One of the known problems with ToF is the so-called *flying points*. These are false points that appear between the edges of the objects and the background. These points can be easily located in the depth image and the 3D point cloud, and easy-to-implement filtering methods are available [3].

Our interest is to place the sensor very close to the scene components, usually in a range from 30 to 50 cm. This high proximity makes ToF cameras easier to calibrate but more susceptible to some error sources. For example, *depth distortion* can be approximated linearly due to the reduced range, and the *built-in pixel errors* can be approximated with a lookup table. Special care should be taken to compensate errors due to saturation (amplitude related) [8], light scattering [52], and multiple light reflections [9]. Note that newer ToF cameras allow easy detection of saturated pixels.

ToF cameras are evolving and a lot of work is being carried out to understand the source of errors and compensate them. The next section presents a classification and short description of the different errors. A detailed ToF error description and classification can be found in [3].

6.4.1 Systematic Errors

Five types of systematic errors have been identified:

Depth distortion appears as a consequence of the fact that the emitted infrared light cannot be generated in practice as theoretically planned (generally sinusoidal) due to irregularities in the modulation process. This type of error produces an offset that depends only on the measured depth for each pixel. Usually, the error plotted against the distance follows a sinusoidal shape[3] (see Figure 6.3). This error is sometimes referred to as *wiggling* or *circular error.*

Built-in pixel-related errors arise from two main sources. On the one hand, errors are due to different material properties in CMOS gates. This produces a constant pixel-related distance offset, leading to different depths measured in two neighbor pixels corresponding to the same real depth. On the other hand, there are latency-related offset errors due to the capacitor charge time delay during the signal correlation process. This can be observed as a rotation on the whole scene (Figure 6.4a) reporting wrong depth measurements. After calibration, the complete scene pose can be correctly recovered (Figure 6.4b).

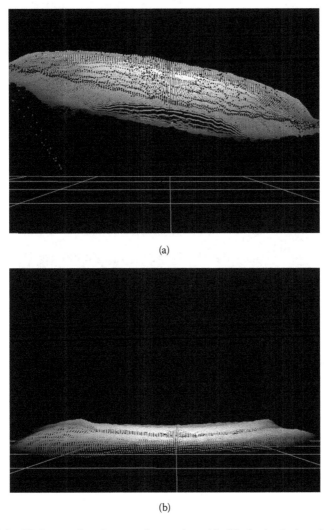

(a)

(b)

FIGURE 6.4   3D image of a planar surface and a white/black checkerboard. (a) Surface should be horizontal, but *built-in pixel-related* error causes a distortion. (b) Once calibrated, the orientation of the plane and the depth of individual points are corrected.   *(continued)*

(c)

(d)

FIGURE 6.4 (CONTINUED)   3D image of a planar surface and a white/black checkerboard. (c) Observe the difference in depth between the squares. (d) The calibrated image is rectified taking into account built-in and amplitude errors.

Amplitude-related errors occur due to low or saturated reflected amplitudes. Low amplitude appears more often in the border of the image, as the emitted light is lower that in the center where the depth is overestimated. Contrarily, when the object is too close to the camera, saturation can appear and depth is underestimated (observe Figure 6.2). Moreover, amplitude-related errors occur due to differences in the object reflectivity, causing differences in the amount of reflected light, and thus yielding different depth measurements for the same constant distance. This effect can be recognized in Figure 6.4c. The image corresponds to a typical calibration pattern: a checkerboard of white and black squares. Observe the difference in depth of the points corresponding to squares of each color.

Both pixel-related errors (depth and amplitude) cause constant depth mismeasurements and can be compensated by means of pixel-offset-based calibration methods like the so-called fixed-pattern noise (FPN) [54]. After the correction determined by calibration, the checkerboard 3D structure can be recovered (Figure 6.4b).

Integration time-related error. Integration time (IT) can be selected by the user. It has been observed that for the same scene different ITs cause different depth values in the entire scene. The main reason for this effect is still the subject of investigation.

Temperature-related errors happen because internal camera temperature affects depth processing, explaining why some cameras include an internal fan. Depth values suffer from a drift in the whole image until the temperature of the camera is stabilized.

6.4.2 Nonsystematic Errors

Four nonsystematic errors can also be identified in depth measurements with ToF cameras, the occurrence of the last three being unpredictable:

Signal-to-noise ratio distortion appears in scenes not uniformly illuminated. Low-illumination areas are more susceptible to noise than high-illumination ones. This type of error is highly dependent on the amplitude, the IT parametrization, and the depth uniformity of the scene. Nonuniform depth over the scene can lead to low-amplitude areas (far objects) that will be highly affected by noise. Usually the IT is calculated to optimally increase accuracy on the distance range of the working scene area.

Multiple light reception errors appear due to the interference of multiple light reflections captured at each sensor's pixel. These multiple light reflections depend on the geometric shape of the objects in the scene and can have two origins. The more obvious is due to concavities that cause multiple reflections. The other one is produced when different depths project to the same pixel; it is more obvious in the edges of the objects and generates the so-called flying points between foreground and background. Flying points can be detected and filtered out (see Figure 6.5).

Light scattering effect arises due to multiple light reflections between the camera lens and its sensor. This effect produces a depth underestimation over the affected pixels, because of the energy gain produced by its neighboring pixel reflections [55]. Errors due to light scattering are only relevant when nearby objects are present in the scene. The closer an object, the higher the interference [56]. These kind of errors are hard to rectify, but some ToF cameras permit the identification of overexposed pixels using some control flags.

Motion blurring, present when traditional cameras are used in dynamic environments, also appears with ToF cameras. This is due to the physical motion of the objects or the camera during the integration time used for sampling.

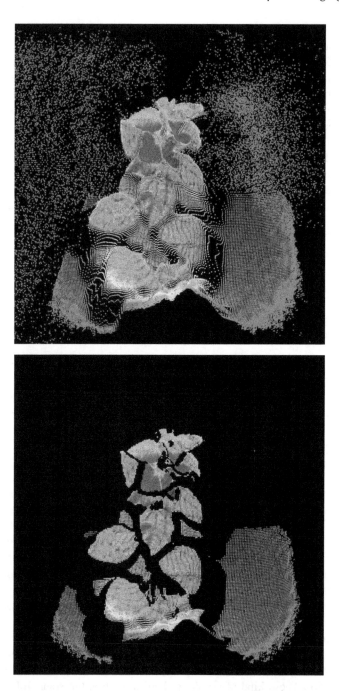

FIGURE 6.5   (See color insert.) Reduction of noise by filtering pixels using a flying-point detector and depth threshold filtering.

6.5 COLORING DEPTH POINTS

The combination of ToF images and color images can be performed to obtain colored point clouds [57], like the ones delivered by Kinect, using the extrinsic calibration parameters between both cameras.

Traditionally extrinsic calibration has been addressed by considering the intensity image of the ToF camera and using classical stereo calibration algorithms [43, 50, 58, 59]. However, the characteristic low resolution of this type of camera leads to a poor localization of the calibration pattern points and the obtained calibration parameters are usually noisy.

Thus, the idea is to take advantage of depth information when calibrating. Once a color camera has been calibrated with a known pattern, reconstruction of the calibration poses is possible, and this can be used to find better extrinsic parameters [60]. A software to calibrate one or multiple color cameras with a ToF camera using this principle is available [61]. This algorithm also includes a depth calibration model that represents the depth deviation as a polynomial function, similar to [9].

Once the extrinsic parameters of the coordinate transformation between a color camera and a ToF camera have been obtained, data fusion is possible. The easy part is to find the correspondences between them and put color to the depth image, but more can be done. Due to the difference in resolution (i.e., 204×204 pixels for a CamCube image, 640×480 for a color image), between each pair of neighbor points in the ToF image there are several points in the color image. As a consequence, these points can be interpolated to obtain a dense depth map [58] where all the color points can be used.

Bartczak et al. [62] used a 3D surface mesh that is rendered into the color camera view as an alternative algorithm to obtain a dense depth map. Huhle et al. [38] presented a completely different approach, where the dense depth map is obtained using a Markov random field (MRF). Depths are represented in the model taking explicitly into account the discontinuities, which are used as a prior to perform the alignment.

As is typical in stereovision, some scene points are seen by one camera but not by the other due to their slightly different viewpoints. Consequently, for some depth points it is impossible to find their corresponding one in the color image. Note that these occlusions appear mainly for close objects—precisely our scenario.

Figure 6.6 is a detail of an image acquired with a Cam-Cube + color camera sensor (Figure 6.7a). In this example, occluded points are detected and colored in red using a Z_buffer approach. First, the point cloud is transformed to the RGB camera reference frame using the extrinsic transformation matrix **F**. Ideally, this leads to 3D points projecting to the corresponding pixel in the color image. In the case of occlusion, only the point that is closer to the camera is stored in the Z_buffer. However, as the ToF camera has a lower resolution than the color camera, it is possible that two 3D points (namely, the foreground and the occluded background points) do not project exactly onto the same color point, so no one is removed. This can lead to a mosaic of foreground-background pixels in the regions where occlusions occur. A neighborhood region can be taken into account to build the Z_buffer, so that the depth of neighbors determines whether occlusions are to be considered.

FIGURE 6.6 (See color insert.) Calibration errors produce bad colored points at the edge of the leaf. Additionally, observe the wrong color assignment in some background points, marked automatically in red, corresponding to the shadow of the leaf. This problem arises as the optical axes of the depth camera and the color camera are not the same and some depth points have no correspondence in the color image. Other sensor combinations, like Kinect, suffer the same problem.

To completely avoid the occlusion problem, the ToF and the color camera optical axes should be the same. This can be accomplished using a beam splitter between the two cameras mounted at 90° [63, 64].

6.6   APPLICATIONS

Some examples of the applicability of ToF cameras in eye-in-hand configurations are presented in this section. Three of the main advantages of actively changing the point of view of a ToF camera are highlighted: the easy acquisition of a 3D structure (that allows straightforward foreground-background segmentation), the ability to acquire accurate views of particular details of a scene, and the ability to disambiguate scenes.

The examples are based on recent experiences mainly in the field of plant phenotyping, and to a lesser extent in that of textile manipulation. In plant phenotyping, a large number of plants have to be monitored, searching for unusual plant responses

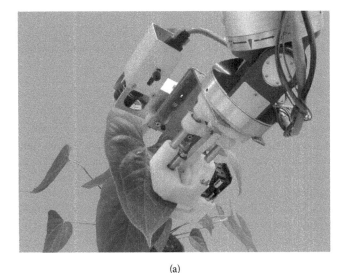

(a)

(b)

FIGURE 6.7   (See color insert.) Details of two different tools in the end effector: (a) a WAM robot and (b) a KUKA lightweight robot. Both tools require that the leaf is placed inside their lateral aperture. An eye-in-hand ToF camera permits acquiring the 3D plant structure required to compute robot motion.

to external factors such as extreme humidity or poor watering. Nowadays, auto-mation of greenhouses provides automatic conveyor belts to transport plants to a measuring cabin, where a set of sensors perform all the measurements required. However, plants can have complex shapes, and having to define the best static posi-tion for all the cameras and other sensors is problematic. The ability to mount a sensor on a manipulator robot in an eye-in-hand configuration is highly appreciated. Additionally, some tasks require placing the sensor or the tool on the surface of a leaf. We provide here two examples of such tasks: the measurement of chlorophyll with a SpadMeter, and the extraction of sample discs for DNA analysis (see both scenarios in Figure 6.7 with the ToF cameras in an eye-in-hand configuration).

6.6.1   3D STRUCTURE

One of the objectives in plant phenotyping is to gather as much information as possible about each specimen, preferably 3D relevant information to enable its subsequent manipulation. Color vision is helpful to extract some relevant features, but it is not well suited for providing the structural/geometric information indispensable for robot interaction with plants. 3D cameras are thus a good complement, since they directly provide depth images. Moreover, plant data acquired from a given viewpoint are often partial due to self-occlusions, and thus planning the best next viewpoint becomes an important requirement. This, together with the need of a high throughput imposed by the application, makes 3D cameras (which provide images at more than 25 fps) a good option in front of other depth measuring procedures, such as stereovision or laser scanners.

Figure 6.8 shows an example of two leaves: the one in the foreground partially occluding the one in the background. Segmentation using only depth information is straightforward. Observe that the background leaf can be better observed after a camera motion. The benefits of moving the camera have some limits in such complex scenarios, as it is not always possible to obtain a better viewpoint, for example, when occlusions are too strong, or when the optimal point of view is out of the working space of the robot.

6.6.2   DETAILED VIEWS

The eye-in-hand configuration allows us to control not only the viewpoint of the camera, but also the distance to the object. To change the distance is also a strategy to change the effective resolution of the image, as relevant details can be better focused.

Figure 6.9 shows the image of a shirt in two different configurations: folded and hanged. Here the task is to grasp the shirt from the collar to allow the robot to hang the shirt in a hanger. Observe that in both configurations, the details of the collar, buttons, and small winkles are visible. In the hanged shirt the sleeves are identifiable as well. Previous works have shown that this 3D structure can be used to identify wrinkles [65] and also the collar structure, using computer vision algorithms [66].

Clearly, the point of view determines the nature of the gathered information, but also the sensor sensitivity determines the relevant details that can be observed. Figure 6.10 shows a view of a plant where the stems are visible. Here, the point of view is important, but also important is that ToF cameras are sensible enough to capture these structures. This is hard to obtain with classical stereovision, and completely impossible with other sensors, like Kinect.

6.6.3   DISAMBIGUATION

Segmentation algorithms use different parameters to adapt to the characteristics of the data, like long ranges, noise type, and sensitivity. The eye-in-hand approach

FIGURE 6.8   (See color insert.) Scene containing a detected leaf occlusion. After changing the point of view the occluded leaf is discovered and more characteristics (e.g., leaf area) can be measured. Depending on the particular leaf arrangement, it is not always possible to completely observe the occluded leaf.

permits moving the camera to find the view that better fits the segmentation parameters. Figure 6.11 shows an example, where in the first view the segmentation algorithm, which uses depth similarity between adjusted surfaces, fails to distinguish two different leaves. Using a next-best-view algorithm [8], a new view is selected that maximizes the difference in depth of the two leaves; thus, the algorithm is now capable of distinguishing the two leaves.

6.7   CONCLUSIONS

ToF cameras have been presented from different perspectives, including underlying principle and characteristics, calibration techniques, applications where camera advantages are explicitly exploited, and potential for future research. Over the last years,

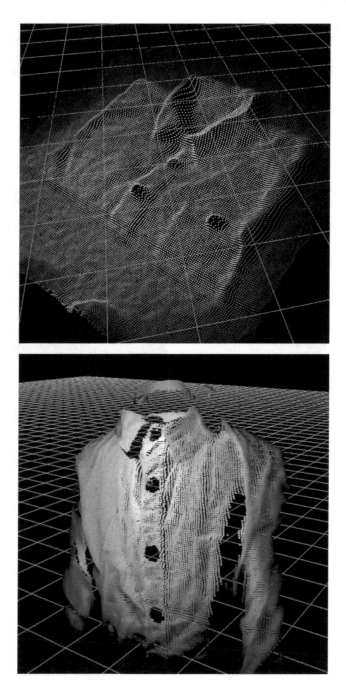

FIGURE 6.9 Details of the perception of a shirt in different configurations. Observe that the small wrinkles are correctly perceived, and some characteristic parts, like the collar shape, are clearly visible.

FIGURE 6.10 (See color insert.) Detail of a plant. Observe that the stems, even if they are thin, are correctly acquired.

FIGURE 6.11 (See color insert.) Scene containing a possible merging of leaves. After changing the point of view, the ambiguity is clarified and two leaves are detected instead of one. Depending on the particular leaf arrangement, it is not always possible to completely disambiguate the occluded leaf.

performance of ToF cameras has improved significantly, errors have been minimized, and higher resolution and frame rates have been obtained. Although ToF cameras cannot yet attain the depth accuracy offered by other types of sensors, such as laser scanners, plenty of research demonstrates that they perform better in many robotic applications. The application of ToF cameras in the wide range of scientific areas we have reviewed indicates their great potential, and widens the horizon of possibilities that were envisaged in the past for vision-based robotics research. We have highlighted here eye-in-hand configurations, where the sensor is mounted on the end effector of a robot manipulator, and it is placed at a short distance from the target object. We have provided experimental evidence of the effectivity of such an approach in three tasks: 3D structure recovering of plants, acquisition of detailed views, and disambiguation.

Advantages of this type of sensor are multiple: they are compact and portable, easing movement; they make data extraction simpler and quicker, reducing power consumption and computational time; and they offer a combination of images that show great potential in the development of data feature extraction, registration, reconstruction, planning, and optimization algorithms, among other positive characteristics. Thus, ToF cameras prove to be especially adequate for eye-in-hand and real-time applications in general, and in particular for automatic acquisition of 3D models requiring sensor movement and on-line mathematical calculation.

Finally, some broad challenges need to be mentioned. First, resolution is still generally low for ToF cameras, despite some efforts having already led to better resolutions, as explained above. Second, short integration times produce a strong noise ratio, and high integration times can result in pixel saturation [10]. Although some algorithms dealing with these problems have already been proposed, more research is needed in this direction. Third, the bistatic configuration (different position of the emitter and the receiver) causes problems in close-range situations because the measured intensity is sensitive to the varying illumination angle. The ability to move the camera is crucial to minimize this effect.

Other concerns include ambient light noise, motion artifacts, and high-reflectivity surfaces in the scene. Ambient light may contain unwanted light of the same wavelength as that of the ToF light source, which may cause false sensor measurements. Frequency-based filters can be used in order to minimize this effect. Motion artifacts are errors caused by receiving light from different depths at the same time due to object motion in the scene. This type of error is mostly observed around the edges of the moving object and can be attenuated either by increasing the frame rate or by correction using motion estimation. Finally, errors due to the coexistence of low-reflective and high-reflective objects (mirroring effect) can be addressed by combining multiple exposure settings.

ACKNOWLEDGMENTS

This work was supported by the Spanish Ministry of Science and Innovation under project PAU+ DPI2011-27510, by the EU Project IntellAct FP7-ICT2009-6-269959, and by the Catalan Research Commission through SGR-00155.

ENDNOTES

1. It is commonly accepted that 0.7 m is the closest distance, but in our tests we have been able to obtain depth images at 0.5 m. The new Kinect camera, to appear in the beginning of 2014, is supposed to work at 0.3 m.
2. Stepfield terrains are the NIST proposal to generate repeatable terrain for evaluating robot mobility.
3. This has been explained as due to perturbations in the measured signal phase caused by aliasing of odd harmonics contained in the emitted reference signal [53].

REFERENCES

1. S. Foix, S. Kriegel, S. Fuchs, G. Alenyà, and C. Torras, Information-gain view planning for free-form object reconstruction with a 3D ToF camera, presented at International

Conference on Advanced Concepts for Intelligent Vision Systems, Brno, Czech Republic, September 2012.

2. G. Alenyà, B. Dellen, S. Foix, and C. Torras, Robotized plant probing: Leaf segmentation utilizing time-of-flight data, *Robotics Autom. Mag.*, 20(3), 50–59, 2013.

3. S. Foix, G. Alenyà, and C. Torras, Lock-in time-of-flight (ToF) cameras: A survey, *IEEE Sensors J.*, 11(9), 1917–1926, 2011.

4. R. Lange and P. Seitz, Solid-state time-of-flight range camera, *IEEE J. Quantum Electron.*, 37(3), 390–397, 2001.

5. PMD-cameras, http://www.pmdtec.com, PMDTechnologies GmbH, 2009.

6. S. F. W. Kazmi and G. Alenyá, Plant leaf imaging using time of flight camera under sunlight, shadow and room conditions, in *IEEE International Symposium on Robotic and Sensors Environments*, Magdeburg, Germany, 2012, pp. 192–197.

7. B. Dellen, G. Alenyà, S. Foix, and C. Torras, 3D object reconstruction from Swissranger sensors data using a spring-mass model, in *Proceedings of the 4th International Conference on Computer Vision Theory and Applications*, Lisbon, February 2009, vol. 2, pp. 368–372.

8. S. Foix, G. Alenyà, J. Andrade-Cetto, and C. Torras, Object modeling using a ToF camera under an uncertainty reduction approach, in *Proceedings of the IEEE International Conference on Robotics and Automation*, Anchorage, AK, May 2010, pp. 1306–1312.

9. S. Fuchs and S. May, Calibration and registration for precise surface reconstruction with time of flight cameras, *Int. J. Int. Syst. Tech. App.*, 5(3–4), 274–284, 2008.

10. J. U. Kuehnle, Z. Xue, M. Stotz, J. M. Zoellner, A. Verl, and R. Dillmann, Grasping in depth maps of time-of-flight cameras, in *Proceedings of the International Workshop on Robotic and Sensors Environments*, Ottawa, ON, October 2008, pp. 132–137.

11. A. Saxena, L. Wong, and A. Y. Ng., Learning grasp strategies with partial shape information, in *Proceedings of the 23rd AAAI Conference on Artificial Intelligence*, Chicago, July 2008, pp. 1491–1494.

12. J. Weingarten, G. Gruener, and R. Siegwart, A state-of-the-art 3D sensor for robot navigation, in *Proceedings of the IEEE/RSJ International Conference on Intelligent Robots and Systems*, Sendei, Japan, September 2004, vol. 3, pp. 2155–2160.

13. S. May, B. Werner, H. Surmann, and K. Pervolz, 3D time-of-flight cameras for mobile robotics, in *Proceedings of the IEEE/RSJ International Conference on Intelligent Robots and Systems*, Beijing, October 2006, pp. 790–795.

14. S. May, K. Pervolz, and H. Surmann, 3D cameras: 3D computer vision of wide scope, in *Vision systems: Applications*. Vienna, Austria: I-Tech Education and Publishing.

15. S. May, D. Droeschel, D. Holz, C. Wiesen, and S. Fuchs, 3D pose estimation and mapping with time-of-flight cameras, in *Proceedings of the IEEE/RSJ IROS Workshop on 3D-Mapping*, Nice, France, September 2008.

16. G. Hedge and C. Ye, Extraction of planar features from Swissranger SR-3000 range images by a clustering method using normalized cuts, in *Proceedings of the IEEE/RSJ International Conference on Intelligent Robots and Systems*, St. Louis, MO, October 2009, pp. 4034–4039.

17. K. Ohno, T. Nomura, and S. Tadokoro, Real-time robot trajectory estimation and 3D map construction using 3D camera, in *Proceedings of the IEEE/RSJ International Conference on Intelligent Robots and Systems*, Beijing, October 2006, pp. 5279–5285.

18. J. Stipes, J. Cole, and J. Humphreys, 4D scan registration with the SR-3000 LIDAR, in *Proceedings of the IEEE International Conference on Robotics and Automation*, Pasadena, CA, May 2008, pp. 2988–2993.

19. S. May, S. Fuchs, D. Droeschel, D. Holz, and A. Nuechter, Robust 3D-mapping with time-of-flight cameras, in *Proceedings of the IEEE/RSJ International Conference on Intelligent Robots and Systems*, St. Louis, MO, October 2009, pp. 1673–1678.

20. P. Gemeiner, P. Jojic, and M. Vincze, Selecting good corners for structure and motion recovery using a time-of-flight camera, in *Proceedings of the IEEE/RSJ International Conference on Intelligent Robots and Systems*, St. Louis, MO, October 2009, pp. 5711–5716.
21. J. Thielemann, G. Breivik, and A. Berge, Pipeline landmark detection for autonomous robot navigation using time-of-flight imagery, in *Proceedings of the IEEE CVPR Workshops*, Anchorage, AK, June 2008, vols. 1–3, pp. 1572–1578.
22. R. Sheh, M. W. Kadous, C. Sammut, and B. Hengst, Extracting terrain features from range images for autonomous random stepfield traversal, in *Proceedings of the IEEE International Workshop on Safety, Security, and Rescue Robotics*, Rome, September 2007, pp. 24–29.
23. A. Swadzba, N. Beuter, J. Schmidt, and G. Sagerer, Tracking objects in 6D for reconstructing static scenes, in *Proceedings of the IEEE CVPR Workshops*, Anchorage, AK, June 2008, vols. 1–3, pp. 1492–1498.
24. S. Acharya, C. Tracey, and A. Rafii, System design of time-of-flight range camera for car park assist and backup application, in *Proceedings of the IEEE CVPR Workshops*, Anchorage, AK, June 2008, vols. 1–3, pp. 1552–1557.
25. O. Gallo, R. Manduchi, and A. Rafii, Robust curb and ramp detection for safe parking using the Canesta ToF camera, in *Proceedings of the IEEE CVPR Workshops*, Anchorage, AK, June 2008, vols. 1–3, pp. 1558–1565.
26. S. B. Goktuk and A. Rafii, An occupant classification system eigen shapes or knowledge-based features, in *Proceedings of the 19th IEEE Conference on Computer Vision and Pattern Recognition*, San Diego, June 2005, pp. 57–57.
27. F. Yuan, A. Swadzba, R. Philippsen, O. Engin, M. Hanheide, and S. Wachsmuth, Laser-based navigation enhanced with 3D time of flight data, in *Proceedings of the IEEE International Conference on Robotics and Automation*, Kobe, Japan, May 2009, pp. 2844–2850.
28. K. D. Kuhnert and M. Stommel, Fusion of stereo-camera and PMD-camera data for real-time suited precise 3D environment reconstruction, in *Proceedings of the IEEE/RSJ International Conference on Intelligent Robots and Systems*, Beijing, October 2006, vols. 1–12, pp. 4780–4785.
29. C. Netramai, M. Oleksandr, C. Joochim, and H. Roth, Motion estimation of a mobile robot using different types of 3D sensors, in *Proceedings of the 4th International Conference on Autonomic and Autonomous Systems*, Gosier, France, March 2008, pp. 148–153.
30. B. Huhle, M. Magnusson, W. Strasser, and A. J. Lilienthal, Registration of colored 3D point clouds with a kernel-based extension to the normal distribution's transform, in *Proceedings of the IEEE International Conference on Robotics and Automation*, Pasadena, CA, May 2008, vols. 1–9, pp. 4025–4030.
31. A. Prusak, O. Melnychuk, H. Roth, I. Schiller, and R. Koch, Pose estimation and map building with a time-of-flight camera for robot navigation, *Int. J. Int. Syst. Tech. App.*, 5(3–4), 355–364, 2008.
32. A. Swadzba, A. Vollmer, M. Hanheide, and S. Wachsmuth, Reducing noise and redundancy in registered range data for planar surface extraction, in *Proceedings of the 19th IAPR International Conference on Pattern Recognition*, Tampa, December 2008, vols. 1–6, pp. 1219–1222.
33. N. Vaskevicius, A. Birk, K. Pathak, and J. Poppinga, Fast detection of polygons in 3D point clouds from noise-prone range sensors, in *Proceedings of the IEEE International Workshop on Safety, Security, and Rescue Robotics*, Rome, September 2007, pp. 30–35.
34. J. Poppinga, N. Vaskevicius, A. Birk, and K. Pathak, Fast plane detection and polygonalization in noisy 3D range images, in *Proceedings of the IEEE/RSJ International Conference on Intelligent Robots and Systems*, Nice, France, September 2008, pp. 3378–3383.

35. K. Pathak, N. Vaskevicius, J. Poppinga, S. Schwertfeger, M. Pfingsthorn, and A. Birk, Fast 3D mapping by matching planes extracted from range sensor point-clouds, in *Proceedings of the IEEE/RSJ International Conference on Intelligent Robots and Systems*, St. Louis, MO, October 2009, pp. 1150–1155.

36. T. Prasad, K. Hartmann, W. Weihs, S. E. Ghobadi, and A. Sluiter, First steps in enhancing 3D vision technique using 2D/3D sensors, in *Computer Vision Winter Workshop*, Prague, February 2006, pp. 82–86.

37. B. Streckel, B. Bartczak, R. Koch, and A. Kolb, Supporting structure from motion with a 3D-range-camera, in *Proceedings of the 15th Scandinavian Conference on Imaging Analysis*, Aalborg, June 2007, vol. 4522, pp. 233–242.

38. B. Huhle, S. Fleck, and A. Schilling, Integrating 3D time-of-flight camera data and high resolution images for 3DTV applications, in *Proceedings of the 1st IEEE International Conference on 3DTV*, Kos Isl, Greece, May 2007, pp. 289–292.

39. S. E. Ghobadi, K. Hartmann, W. Weihs, C. Netramai, O. Loffeld, and H. Roth, Detection and classification of moving objects—Stereo or time-of-flight images, in *Proceedings of the International Conference on Computational Intelligence and Security*, Guangzhou, China, November 2006, vol. 1, pp. 11–16.

40. S. Hussmann and T. Liepert, Robot vision system based on a 3D-ToF camera, in *Proceedings of the 24th IEEE Instrumentation and Measurement Technology Conference*, Warsaw, May 2007, vols. 1–5, pp. 1405–1409.

41. S. A. Guomundsson, R. Larsen, and B. K. Ersboll, Robust pose estimation using the Swissranger SR-3000 camera, in *Proceedings of the 15th Scandinavian Conference on Imaging Analysis*, Aalborg, June 2007, vol. 4522, pp. 968–975.

42. C. Beder, B. Bartczak, and R. Koch, A comparison of PMD-cameras and stereo-vision for the task of surface reconstruction using patchlets, in *Proceedings of the 21st IEEE Conference on Computer Vision and Pattern Recognition*, Minneapolis, MN, June 2007, vols. 1–8, pp. 2692–2699.

43. T. Grundmann, Z. Xue, J. Kuehnle, R. Eidenberger, S. Ruehl, A. Verl, R. D. Zoellner, J. M. Zoellner, and R. Dillmann, Integration of 6D object localization and obstacle detection for collision free robotic manipulation, in *Proceedings of the IEEE/SICE International Symposium on System Integration*, Nagoya, Japan, December 2008, pp. 66–71.

44. U. Reiser and J. Kubacki, Using a 3D time-of-flight range camera for visual tracking, in *Proceedings of the 6th IFAC/EURON Symposium on Intelligent Autonomous Vehicles*, Toulouse, France, September 2007.

45. S. Gächter, A. Harati, and R. Siegwart, Incremental object part detection toward object classification in a sequence of noisy range images, in *Proceedings of the IEEE International Conference on Robotics and Automation*, Pasadena, CA, May 2008, vols. 1–9, pp. 4037–4042.

46. J. Shin, S. Gachter, A. Harati, C. Pradalier, and R. Siegwart, Object classification based on a geometric grammar with a range camera, in *Proceedings of the IEEE International Conference on Robotics and Automation*, Kobe, Japan, May 2009, pp. 2443–2448.

47. Z. Marton, R. Rusu, D. Jain, U. Klank, and M. Beetz, Probabilistic categorization of kitchen objects in table settings with a composite sensor, in *Proceedings of the IEEE/RSJ International Conference on Intelligent Robots and Systems*, St. Louis, MO, October 2009, pp. 4777–4784.

48. J. Zhu, L. Wang, R. Yang, and J. Davis, Fusion of time-of-flight depth and stereo for high accuracy depth maps, in *Proceedings of the 22nd IEEE Conference on Computer Vision and Pattern Recognition*, Anchorage, AK, June 2008, vols. 1–12, pp. 3262–3269.

49. M. Lindner, M. Lambers, and A. Kolb, Sub-pixel data fusion and edge-enhanced distance refinement for 2D/3D, *Int. J. Int. Syst. Tech. App.*, 5(3–4), 344–354, 2008.

50. J. Fischer, B. Huhle, and A. Schilling, Using time-of-flight range data for occlusion handling in augmented reality, in *Proceedings of the Eurographics Symposium on Virtual Environments*, September 2007, pp. 109–116.
51. H. Surmann, A. Nüchter, and J. Hertzberg, An autonomous mobile robot with a 3D laser range finder for 3D exploration and digitalization of indoor environments, *Robotic Auton. Syst.*, 45(3–4), 181–198, 2003.
52. J. Mure-Dubois and H. Hügli, Real-time scattering compensation for time-of-flight camera, presented at *Proceedings of the 5th International Conference on Computer Vision Systems,* Bielefeld, Germany, March 2007.
53. R. Lange, 3D time-of-flight distance measurement with custom solid-state image sensors in CMOS/CCD-technology, PhD dissertation, University of Siegen, Siegen, Germany, 2000.
54. T. Kahlmann, F. Remondino, and H. Ingensand, Calibration for increased accuracy of the range imaging camera Swissranger™, in *ISPRS Commission V Symposium*, Dresden, September 2006, pp. 136–141.
55. T. Kahlmann and H. Ingensand, Calibration and development for increased accuracy of 3D range imaging cameras, *J. Appl. Geodesy*, 2(1), 1–11, 2008.
56. W. Karel, Integrated range camera calibration using image sequences from hand-held operation, in *Proceedings of the ISPRS Conference*, Beijing, July 2008, vol. 37, pp. 945–952.
57. G. Alenyà, B. Dellen, and C. Torras, 3D modelling of leaves from color and ToF data for robotized plant measuring, in *Proceedings of the IEEE International Conference on Robotics and Automation*, Shanghai, May 2011, pp. 3408–3414.
58. M. Lindner, A. Kolb, and K. Hartmann, Data-fusion of PMD-based distance-information and high-resolution RGB-images, in *Proceedings of the International Symposium on Signals Circuits Systems*, Lasi, July 2007, vols. 1–2, pp. 121–124.
59. S. A. Guomundsson, H. Aanæs, and R. Larsen, Fusion of stereo vision and time-of-flight imaging for improved 3D estimation, *Int. J. Int. Syst. Tech. App.*, 5(3–4), 425–433, 2008.
60. I. Schiller, C. Beder, and R. Koch, Calibration of a PMD camera using a planar calibration object together with a multi-camera setup, in *Proceedings of the ISPRS Conference*, Beijing, July 2008, vol. 37, part B3a, pp. 297–302.
61. http://mip.informatik.uni-kiel.de, 2009.
62. B. Bartczak, I. Schiller, C. Beder, and R. Koch, Integration of a time-of-flight camera into a mixed reality system for handling dynamic scenes, moving viewpoints and occlusions in real-time, in *Proceedings of the International Symposium on 3D Data Processing, Visualization and Transmission (3DPVT)*, Atlanta, GA, June 2008.
63. S. Ghobadi, O. Loepprich, F. Ahmadov, and J. Bernshausen, Real time hand based robot control using multimodal images, *Int. J. Comput. Sci.*, 35(4), 2008.
64. J.-H. Cho, S.-Y. Kim, Y.-S. Ho, and K. Lee, Dynamic 3D human actor generation method using a time-of-flight depth camera, *IEEE Trans. Consumer Electron.*, 54(4), 1514–1521, 2008.
65. A. Ramisa, G. Alenyà, F. Moreno-Noguer, and C. Torras, Determining where to grasp cloth using depth information, in *Proceedings of the 14th International Conference of the Catalan Association for Artificial Intelligence*, Lleida, Spain, October 2011.
66. A. Ramisa, G. Alenyá, F. Moreno-Noguer, and C. Torras, Using depth and appearance features for informed robot grasping of highly wrinkled clothes, in *Proceedings of the IEEE International Conference on Robotics and Automation*, St. Paul, MN, May 2012, pp. 1703–1708.

7 Smart Eye-Tracking Sensors Based on Pixel-Level Image Processing Circuits

Dongsoo Kim

CONTENTS

7.1 INTRODUCTION AND GENERAL ISSUES IN EYE TRACKERS

The common pointing devices in a computer system are a mouse and a pen digitizer. However, these interfaces are difficult to be use in mobile environments wearing a head-mounted display (HMD). One of the alternative pointing devices is an eye tracker that acquires the point on the screen on which the user gazes [1–3]. Figure 7.1 shows the eye tracker in HMD. The eye tracker gives the information to HMD that indicates the point on which the user gazes.

Infrared light is commonly used in the eye tracker because it eliminates the influence of ambient illumination and improves the discrepancy between the pupil and the white of the eye [4]. Under infrared illumination, the pupil is the biggest black region in the eye image. Therefore, the point that the eye gazes on can be obtained by finding the center point of the pupil in the eye image.

A common eye tracker uses a normal image sensor and image processing algorithm [3–5]. However, this system needs an analog-to-digital converter (ADC), interface circuit, and software computation. As the tracking resolution is increased, the tracking time is increased drastically due to the speed limitation of ADC, data interface, and computation. To overcome these disadvantages, smart image sensors that include image processing and pattern recognition circuits on the same silicon chip have been reported in [6–11]. This chapter introduces three versions of the single-chip eye tracker with the smart complementary metal-oxide-semiconductor (CMOS) image sensor pixels.

7.1.1 EYE MOVEMENT

The oculomotor system can be analyzed more easily than most other movement control systems because it can be broken down functionally into smaller subsystems [12], which can be analyzed individually. Eye movements can generally be grouped into six types: saccades, smooth pursuit, vergence, optokinetic reflex, vestibulo-ocular reflex, and fixation. Table 7.1 demonstrates the velocity range and angular range of these eye movements.

Saccades are the rapid eye movements when we look around to examine our environment. Saccadic eye movements can vary considerably in magnitude, from less than

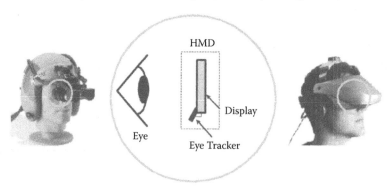

FIGURE 7.1 Eye tracker in HMD.

TABLE 7.1

Velocity and Angular Ranges for Various Eye Movements

Type of Movement	Velocity Range	Angular Range
Saccade	10–800°/s (2–170 mm/s)	0.1–100°
Smooth pursuit and optokinetic reflex	<40°/s (8 mm/s)	>1°
Vergence	10°/s (2 mm/s)	~15°
Vestibulo-ocular reflex	100–400°/s (21–84 mm/s)	1–12°
Fixation	10–300°/s (0.03–1 mm/s)	0.003–0.2°

1° to more 100° as the situation demands. The peak velocity during the movement depends on the size of the saccade. For large saccades, it can reach 800°/s; because the velocity is so high, the movements are brief. The most frequently occurring saccades, which are smaller than 15°, take less than 50 ms from start to finish. Saccades are voluntary eye movements in the sense that we can consciously choose to look at, or ignore, things in the visual scene. Once initiated, however, the execution of the movement is automatic. During the eye movement, when the retinal image is sweeping rapidly across the retina, vision is actively suppressed.

Smooth pursuit movements are visually guided movements in which the eye tracks a small object that is moving relative to a stationary background. The purpose of smooth pursuit movements is to stabilize the retinal image of a moving object in order to allow it to be visually examined. Pursuit movements are voluntary only in so far as we can choose to look at a moving object or ignore it. The kinematics of pursuit movements are completely determined by the motion of the object that is being tracked. It has the peak velocity of 40°/s.

Vergence moves the eyes in opposite directions, causing the intersection point of the lines of sight of the two eyes to move closer or farther away. The purpose of vergence eye movements is to direct the fovea (area of the retina with the highest resolution) of both eyes at the same object. The vergence movements have an angular range of about 15°. Under natural conditions vergence movements are accompanished by saccades or pursuit movements.

The optokinetic reflex is a visually guided reflex, the purpose of which is to compensate body and head movements so that retinal image motion is minimized. The optokinetic reflex responds optimally if the stimulus is movement of all or a large portion of the retinal image. We have no voluntary control over these reflexive eye movements. If the retinal image of the whole field of view moves, our eyes invariably follow this mothion. Just as with pursuit movements, the kinematics of eye movements resulting from the optokinetic reflex are determined by the motion of the stimulus.

The vestibulo-ocular reflex is similar to the optokinetic reflex in that its function is also to compensate for head motion and stabilize the retinal image. In contrast to the optokinetic reflex, however, the vestibulo-ocular reflex is not guided visually, but is generated by receptors in the inner ear that detect acceleration and changes with respect to gravity. The vestibulo-ocular reflex is also beyond our voluntary control. The kinematics of these eye movements are directly linked to the signals that are generated by the vestibular organs in the inner ear. The vestibular movements have a velocity range from 100 to 400°/s.

Fixation or visual fixation is the maintaining of the visual gaze on a location. Humans (and other animals with a fovea) typically alternate saccades and visual fixations. Visual fixation is never perfectly steady: this eye movement occurs involuntarily. The term *fixation* can also be used to refer to the point in time and space of focus rather than to the act of fixating; a fixation in this sense is the point between any two saccades, during which the eyes are relatively stationary and virtually all visual input occurs. The fixation always lasts at least 100 ms. It is during these fixations that most visual information is acquired and processed. At least three types of small involuntary eye motions commonly occur during the fixations. Flicks are very rapid (perhaps as little as 30 ms apart), involuntary, saccade-like motions of less than 1°. Drifts are very small and slow (about 0.1°/s). They are apparently random motions of the eye. Tremors are tiny (less than 30 arcsec), high-frequency (30–150 Hz) eye vibrations [13].

7.1.2   EYE-TRACKING TECHNIQUE

There exist four main methods for measuring eye movements in relation to head movements: electro-oculography (EOG), scleral contact lens/search coil, photo-oculography (POG) or video-oculography (VOG), and video-based combined pupil and corneal reflection [14].

7.1.2.1   Electro-Oculography

Electro-oculography is based on the electronic measurement of the potential created by differences between the cornea and the retina when the eye is rotated, and is the most widely applied eye movement recording method. A picture of a subject wearing the EOG apparatus is shown in Figure 7.2. The recorded potentials exist in the range of 15–200 μV, with nominal sensitivities of the order of 20 μV/degree of eye movement. This technique measures eye movements relative to head position, and so is not generally suitable for point-of-regard measurements unless the head position is also measured.

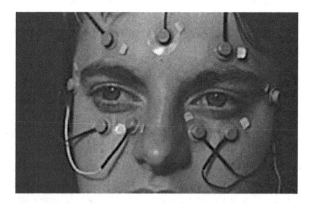

FIGURE 7.2   Example of EOG eye movement measurement. (From Metrovision, France, http://www.metrovision.fr. Copyright © Metrovision.)

7.1.2.2   Scleral Contact Lens/Search Coil

Scleral contact lenses/search coils involve attaching a mechanical or optical object to a large contact lens that is worn directly on the eye (covering both the cornea and scleral). Figure 7.3 shows a picture of the search coil embedded in a scleral contact lens and the electromagnetic field frame. Although the scleral search coil is the most precise eye movement measurement method, it is also the most intrusive. Insertion of the lens requires care and practice. Wearing of the lens causes discomfort. This method also measures eye position relative to the head, and is not generally suitable for point-of-regard measurement.

7.1.2.3   Photo-Oculography or Video-Oculography

Photo-oculography or video-oculography groups together a wide variety of eye movement recording techniques that involve the measurement of distinguishable features of the eyes under rotation/translation, such as the apparent shape of the pupil, the position of the limbus, and corneal reflections of a closely situated directed light source. Automatic limbus tracking often uses photodiodes mounted on spectable frames, as shown in Figure 7.4. This method requires the head to be fixed, e.g., by using either a head/chin rest or a bite bar.

7.1.2.4   Video-Based Combined Pupil-Corneal Reflection

Although the above techniques are in general suitable for eye movement measurements, they often do not provide point-of-regard measurement. Video-based combined pupil and corneal reflection is an eye movement measurement technique that is able to provide information about where the user is looking in space, while taking the user's eye positions relative to the head into account. Video-based trackers utilize inexpensive cameras and image processing hardware to compute the point of regard in real time. The apparatus may be table mounted, as shown in Figure 7.5, or worn on the head, as shown in Figure 7.6. The optics of both table-mounted and head-mounted systems are essentially identical, with the exception of size. These devices, which are becoming increasingly available, are most suitable for use in interactive systems.

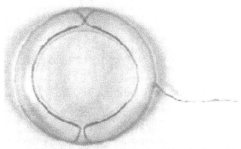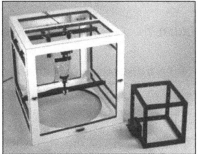

FIGURE 7.3   Example of search coil embedded in contact lens and electromagnetic field frames for search coil eye movement measurement. (From Skalar Medical, Netherlands, http://www.skalar.nl. Copyright © Skalar Medical.)

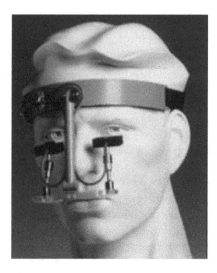

FIGURE 7.4   Example of infrared limbus tracker apparatus. (From Microguide, USA, http://www.eyemove.com. Copyright © Microguide.)

FIGURE 7.5   Example of table-mounted video-based eye tracker.

7.1.3   ILLUMINATION OF THE EYE TRACKER

Figure 7.7 shows an eye image under visual and infrared illumination. An iris is the biggest black region under visual illumination, and its region is covered by the eyelid. The boundary between the iris and the white of the eye is not distinct. However, under infrared illumination the pupil is the biggest black region, and it is a circle with a clear boundary. Infrared illumination eliminates the influence of ambient illumination. Therefore, a common eye tracker uses infrared illumination to find the center of the eyeball. A diffuser can be used to spread out the bright spot caused by the infrared light source, and a contact lens does not have a significant effect on the eye tracker.

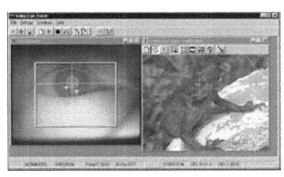 

FIGURE 7.6 Example of head-mounted video-based eye tracker. (From IOTA AB, EyeTrace Systems, Sundsvall Business and Technology, http://www.iota.se. Copyright © IOTA AB.)

7.1.4 EVALUATION CRITERION FOR THE EYE TRACKER

In this section we discuss some of the evaluation criterion for the eye-tracking system. The performance of the eye tracker can be evaluated by the resolution, speed (sampling rate), transport delay, accuracy, and other considerations. Table 7.2 lists several general

---

TABLE 7.2

Performance Parameters for General Applications

Measure point of regard or scan path for off-line analysis
- Accuracy = 1° visual angle
- Sample rate = 60 Hz

Real-time control using point of regard (switch selection, display control)
- Transport delay ≤ 50 ms
- Sample rate = 60 Hz
- Accuracy = 1°

Study saccadic velocity profiles, nystagmus
- Sample rate ≥ 240 Hz
- Linearity ≤ 5%
- Resolution = 0.25°

Study flicks, drifts
- Sample rate ≥ 240 Hz
- Linearity ≤ 5%
- Resolution = 10 arcmin

Measure tremor
- Sample rate ≥ 1000 Hz
- Linearity ≤ 5%
- Resolution = 1 arcsec

Stabilize image on retina
- Accuracy = 2 arcmin
- Sample rate ≥ 240 Hz
- Transport delay ≤ 10 ms

---

application categories, the performance parameters that are probably most important, and values for those parameters that are probably in an appropriate range.

7.1.5 CELLULAR NEURAL NETWORKS

The basic circuit unit of cellular neural networks is called a *cell*. It contains linear and nonlinear circuit elements, which typically are linear capacitors, linear resistors, linear and nonlinear controlled sources, and independent sources. The structure of cellular neural networks is similar to that found in cellular automata; namely, any cell in a cellular neural network is connected only to its neighbor cells. The adjacent cells can interact directly with each other. Cells not directly connected together may affect each other indirectly because of the propagation effects of the continuous-time dynamics of cellular neural networks. An example of a two-dimensional cellular neural network is shown in Figure 7.8.

A typical example of a cell C_{ij} of a cellular neural network is shown in Figure 7.9, where the suffices u, x, and y denote the *input*, *state*, and *output*, respectively. The

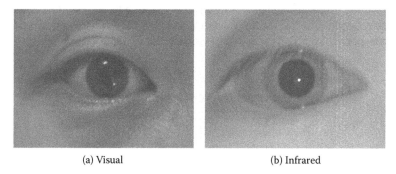

(a) Visual   (b) Infrared

FIGURE 7.7   Eye image under visual and infrared illumination.

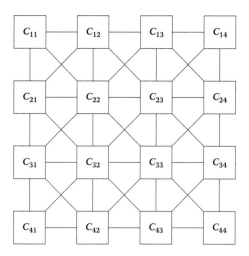

FIGURE 7.8   Two-dimensional cellular neural network.

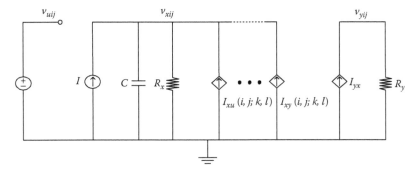

FIGURE 7.9   Example of a cell circuit.

node voltage v_{xij} of C_{ij} is called the state of the cell, and its initial condition is assumed to have a magnitude of less than or equal to 1. The node voltage v_{uij} is called the input of C_{ij} and is assumed to be a constant with a magnitude less than or equal to 1. The node voltage v_{yij} is called the output.

Observe from Figure 7.9 that each cell C_{ij} contains one independent voltage source E_{ij}, one independent current source I, one linear capacitor C, two linear resistors R_x and R_y, and at most $2m$ linear voltage-controlled current sources, which are coupled to its neighbor cells via the controlling input voltage v_{ukl}, and the feedback from the output voltage v_{ykl} of each neighbor cell C_{kl}, where m is equal to the number of neighbor cells. In particular, $I_{xy}(i, j; k, l)$ and $I_{xu}(i, j; kl)$ are linear voltage-controlled current sources with the characteristics $I_{xy}(i, j; kl) = A(i, j; kl)v_{ykl}$ and $I_{xu}(i, j; kl) = B(i, j; kl)v_{ukl}$ for all neighbor cells C_{kl}.

Applying KCL and KVL, the circuit equations of a cell with an $M \times N$ cell array are easily derived, as follows:

State equation:

$$C\frac{dv_{xij}(t)}{dt} = -\frac{1}{R_x}v_{xij}(t) \; + \sum_{C_{kl};\,neighbor\,cell\,of\,C_{ij}} A(i,j;k,l)\,v_{ykl}(t)$$

$$+ \sum_{C_{kl};\,neighbor\,cell\,of\,C_{ij}} B(i,j;k,l)\,v_{ukl} + I$$

$$1 \le i \le M; 1 \le j \le N$$

Output equation:

$$v_{yij}(t) = \frac{1}{2}\left(\left|v_{xij}(t) + 1\right| - \left|v_{xij}(t) - 1\right|\right), \quad 1 \le i \le M; 1 \le j \le N$$

Input equation:

$$v_{uij} = E_{ij}, \quad 1 \le i \le M; 1 \le j \le N$$

Constraint conditions:

$$\left|v_{xij}(0)\right| \le 1, \quad 1 \le i \le M; 1 \le j \le N$$

$$\left|v_{uij}\right| \le 1, \quad 1 \le i \le M; 1 \le j \le N$$

Parameter assumptions:

$$A(i,j;k,l) = A(k,l;i,j), 1 \le i, k \le M; 1 \le j, l \le N$$

$$C > 0, R_x > 0$$

7.1.6   EYE TRACKER CALIBRATION

When the eye tracker interacts with the display of HMD or the graphical applications, the most important requirement is mapping of eye tracker coordinates to the
display image coordinates.

In the monocular case, a typical 2D image-viewing application is expected, and
the coordinates of the eye tracker are mapped to the 2D (orthogonal) viewport coordinates accordingly (the viewport coordinates are expected to match the dimensions
of the image displayed in HMD). In the binocular (VR) case, the eye tracker coordinates are mapped to the dimensions of the near viewing plane of the viewing frus-
tum. The following sections discuss the mapping and calibration of the eye tracker
coordinates to the display coordinates for the monocular applications.

For the calibration of the eye tracker, the data obtained from the tracker must be
mapped to a range appropriate to the 2D display image coordinates, as shown in
Figure 7.10 [14].

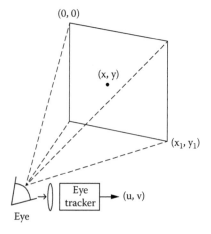

FIGURE 7.10   Calibration method of the eye tracker.

A linear mapping between two coordinate systems can be adopted. In general, if $u \in [u_0, u_1]$ needs to be mapped to the range $[0, x_1]$, we have

$$x = (u - u_0) \cdot x_1/(u_1 - u_0) \tag{7.1}$$

Equation (7.1) has a straightforward interpretation:

1. The value u is translated to its origin, by subtracting u_0.
2. The value $(u - u_0)$ is then normalized by dividing by the range $(u_1 - u_0)$.
3. The normalized value, $(u - u_0)/(u_1 - u_0)$, is then scaled to the new range (x_1).

The y coordinate can also be obtained by the same linear mapping as follows:

$$y = (v - v_0) \cdot y_1/(v_1 - v_0) \tag{7.2}$$

To reduce the mapping error, a bilinear transform can be used. If the coordinates (u, v) of the eye tracker have a trapezoid shape, as shown in Figure 7.11, then u_{01}, u_{32}, v_{03}, and v_{12} can be obtained as

$$u_{01} = u_0 + x(u_1 - u_0)$$
$$u_{32} = u_0 + x(u_2 - u_3) \tag{7.3}$$
$$v_{03} = v_0 + y(v_3 - v_0)$$

Then (u, v) can be calculated as follows:

$$u = u_{01} + y(u_{32} - u_{01}) \tag{7.4}$$
$$v = v_{03} + x(v_{12} - v_{03})$$

The final coordinates (x, y) can be solved by the mathematical solving tools [20, 21] from (7.3) and (7.4).

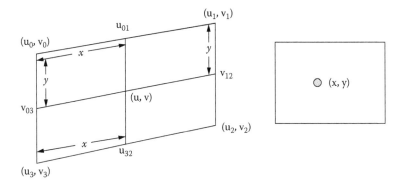

FIGURE 7.11   Bilinear transform for the calibration of the eye tracker.

7.2   FIRST VERSION OF THE EYE TRACKER

This section proposes the first version of the single-chip eye tracker that generates the address for the center point of the pupil. Section 7.2.1 proposes the eye tracker architecture and operation principle. Section 7.2.2 describes circuit implementation. Section 7.2.3 presents the simulation and experimental results. Finally, Section 7.2.4 provides the conclusion.

7.2.1   PROPOSED EYE TRACKER ARCHITECTURE

The proposed eye tracker is composed of a smart CIS pixel array, winner-take-all (WTA) circuits, address encoders, and a stop criterion circuit, as shown in Figure 7.12. The pixel array includes a photosensor and interpixel feedback circuit that captures the image and performs a shrink opération, which is described below. The WTA circuit finds the winning current and the location from the horizontally/vertically summed current from the pixel array. The winning location is encoded by the encoder and latched by the trigger signal, V_{STOP}, from the stop criterion circuit.

Figure 7.13 shows the flowchart of the operation principle that is explained below. The image is captured by photodiode and then transferred as an initial state voltage. During the shrink phase the black region of the image is shrunk by interpixel positive feedback. Each pixel generates current, I_{ij}, that is idnversely proportional to its state voltage, and these currents are summed column-wise and row-wise, respectively. These summed currents (I_i, I_j) represent how many black pixels are in each column

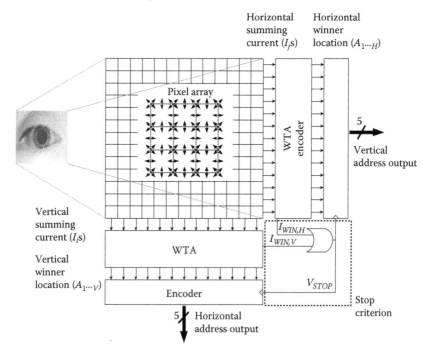

FIGURE 7.12   Block diagram of the proposed eye tracker. (Copyright © Kluwer 2005.)

and row, respectively. The WTA circuit detects the column and row that has the largest number of black pixels during the shrink operation. Operations 1, 2, 3, and 4 in
Figure 7.13 are performed simultaneously by the analog circuits. This procedure is
continuously repeated until the number of black pixels is less than 2. When the number of black pixels is less than 2, the winning column and row address are latched
and read out. These addresses indicate coordinates of the remaining black pixels that
correspond to the center of the pupil. Once the address is read out, a new image is
captured and the above procedure is repeated with the new image.

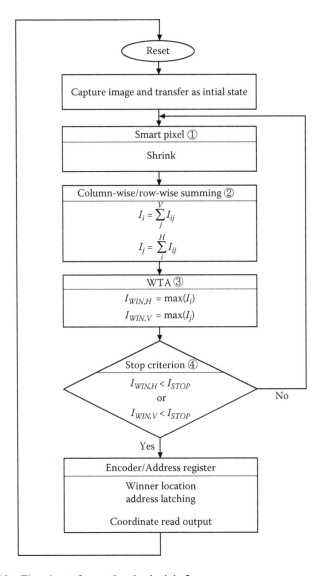

FIGURE 7.13   Flowchart of operational principle 5.

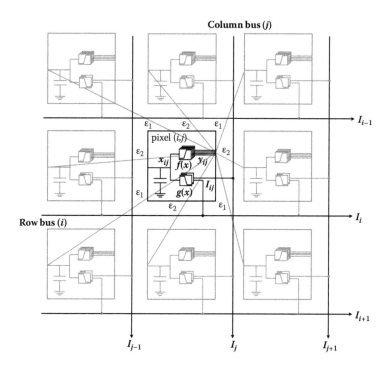

FIGURE 7.14   Block diagram of the smart CIS pixel array. (Copyright © Kluwer 2005.)

The smart CIS pixel array performs shrink operation with continuous-time inter-pixel feedback, as shown in Figure 7.14. The dynamics of the proposed smart pixel can be expressed as

$$
\left\{
\begin{aligned}
& C_F \frac{dx_{ij}}{dt} = \sum_{k=-1}^{1}\sum_{l=-1}^{1} \varepsilon_{kl} y_{(i+k,j+l)}, \; x_{ij}(0) = u_{ij} \\
& y_{ij} = f(x_{ij})
\end{aligned}
\right.
\tag{7.5}
$$

where C_F is the integration capacitor and u_{ij} is the initial state of the pixel (i, j) that is captured from the photosensor. x_{ij} and y_{ij} represent the state voltage and output current of pixel (i, j), respectively. $f(x_{ij})$ is a transconductor with saturation, and ε_{kl} is the interpixel feedback gain coefficient, as shown in Figure 7.12. Here, the feedback coefficients should satisfy $0 < \varepsilon_1 < \varepsilon_2 < 1$, and the self-feedback should be zero for the isotropic shrink operation.

During the shrink operation, the current I_{ij} is generated by each pixel as $I_{ij} = g(x_{ij})$. I_{ij} is generated when the pixel is considered a black pixel and is inversely proportional to x_{ij}. The currents I_{ij} are summed column-wise and row-wise, as shown Figure 7.14, and denoted as I_i and I_j, where i and j are row number and column number, respectively. Since the black pixel (low x_{ij}) generates a high I_{ij}, the summed currents represent how many black pixels remain in each column and row.

Since each pixel gives positive feedback to its surrounding pixels, a white pixel forces its neighbor pixels to be white, while a dark pixel provides a small amount of feedback to its neighbor pixels. This interpixel feedback causes the boundary's black pixels to be changed to white continuously, and the size of the black region is shrunk, as shown in Figure 7.15. The nonlinear function $f(x_{ij})$ causes the gray image to be

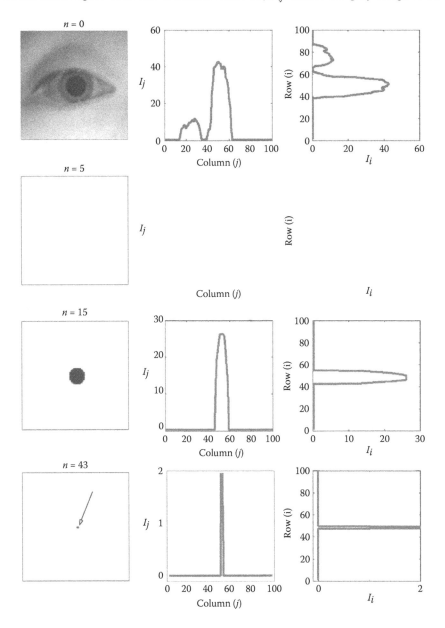

FIGURE 7.15   System level simulation results of shrink operation. (Copyright © Kluwer 2005.)

gradually changed to the black and white image. This shrink operation continues and eventually all pixels become white.

Figure 7.16 shows the block diagram of the WTA, stop criterion circuit, address encoder, and address latch. Each WTA continuously detects the column/row that has the highest summed current (I_i or I_j) during the shrink operation. The WTA generates the winner location logic output $A_1, \ldots$, and only one node out of $A_1, \ldots$ is high logic and represents the winning column/row. The WTA circuit not only identifies the location of the winner, but also selects the winning current, I_{WIN}. I_{WIN} is used by the stop criterion circuit to decide whether the shrink operation is completed.

The address encoder continuously encodes the winning row/column during the shrink operation. If the wining current (I_{WIN}) is smaller than a certain limit (I_{STOP}) that corresponds to two black pixels, it means that the largest black region is completely shrunk down to 2 pixels, located at the center of the largest black object. In this situation, the winner location logic output represents the location of the center of the remaining pixels. The stop criterion circuit generates a high logic signal when either a horizontal or vertical winning current reaches I_{STOP}, and it causes the address registers to latch the winning address from the encoders. This latched address corresponds to the center of the pupil.

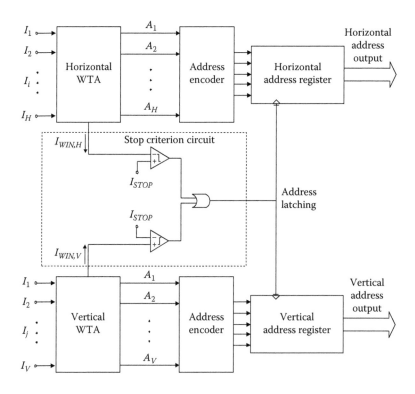

FIGURE 7.16 Block diagram of WTA, address encoder, and stop criterion. (Copyright © Kluwer 2005.)

7.2.2 Circuit Implementation

The schematic diagram of the proposed smart CIS pixel is shown in Figure 7.17a. Figure 7.17b represents the timing diagram for one frame. First, the photodiode (PD) capacitor, C_S is precharged to V_{REF1} and C_F is reset during the *reset* phase by turning on the *TX* and *RST* switches. Second, the *TX* switch is turned off and C_S is discharged corresponding to the incident light during the *capture* phase. In the *transfer* phase, the charge in C_S is transferred to C_F by turning on the *TX* switch and turning off the *RST* switch. The C_S is again reset to V_{REF1} at this moment. Once the PD charge is transferred into C_F, x_{ij} represents the light intensity (u_{ij}) of the pixel (i, j).

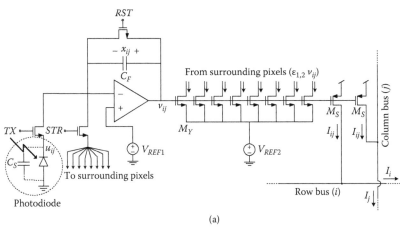

(a)

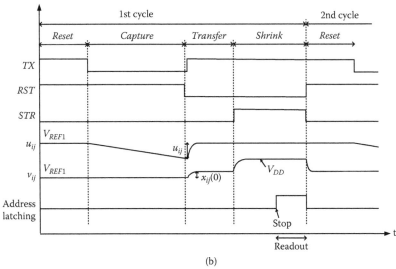

(b)

FIGURE 7.17   Block and timing diagrams of the smart CIS pixel. (a) Schematic diagram of the smart CIS pixel; (b) Timing diagram of the proposed eye tracker. (Copyright © Kluwer 2005.)

The transistors M_Y generate feedback currents, εy_{ij}, corresponding to the pixel state voltage x_{ij}. In *shrink* phases, the *STR* switch is turned on and the currents to the surrounding pixels are summed and integrated in C_F, causing x_{ij} to be increased.

Here, C_F is chosen four times larger than C_S because C_S is in the range of several tens of femtofarad, and the use of such a small capacitance for C_F causes high sensitivity to parasitic capacitance. This ratio causes x_{ij} to be one-fourth of u_{ij}. Figure 7.18 shows implementation of $f(x_{ij})$ and $g(x_{ij})$ by careful arrangement of V_{REF1} and V_{REF2} considering the operating region of M_Y and M_S. Since the dynamic range of u_{ij} is bounded as $0 < u_{ij} < V_{REF1}$ and $x_{ij} = 1/4u_{ij}$, V_{REF1} is chosen about $4/5V_{DD}$ considering the dynamic range and the implementation of the saturation function $f(x_{ij})$. The V_{REF2} is chosen so that M_Y is operated to cutoff when the x_{ij} is low. The $g(x_{ij})$ is realized by M_S's threshold voltage. M_S is turned on when x_{ij} is sufficiently low to be considered a black pixel.

Figure 7.19 presents a schematic diagram of the WTA [22–25] and stop criterion circuit. It is decided to turn on only the common source voltage, V_S, the winning

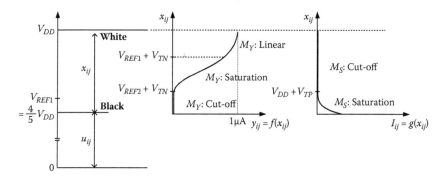

FIGURE 7.18 Implementation of $f(x_{ij})$ and $g(x_{ij})$. (Copyright © Kluwer 2005.)

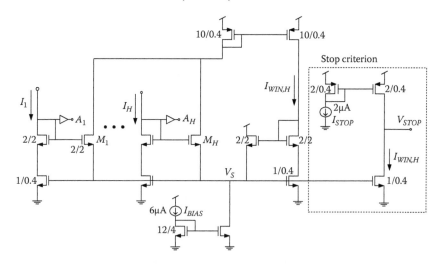

FIGURE 7.19 Circuit diagram of the WTA circuit and stop criterion. (Copyright © Kluwer 2005.)

transistor out of $M_{1...H}$ that has the highest input current, and all others are turned off. The bias current, I_{BIAS}, is chosen to be twice the maximum I_{ij} generated by one black pixel. The winner current, I_{WIN}, is equal to the highest input current and is copied to the stop criterion circuit.

Since only the winning transistor is turned on and allows current flow (Figure 7.20b), only the winner generates high logic, winning location logic output ($A_{1...H}$), and all others are low. The $A_{1...H}$ outputs are connected to the address encoder. Although the WTA deals with a large number of inputs, the mismatch among transistors is not critical because eventually the WTA identifies the winner from two adjacent inputs that correspond to the remaining two black pixels in the shrink operation. In the stop criterion circuit, the address latching signal, V_{STOP}, is generated by comparing I_{WIN} and I_{STOP}. The V_{STOP} is changed to high logic when I_{WIN} is smaller than I_{STOP}.

7.2.3 SIMULATION AND EXPERIMENTAL RESULTS

Figure 7.20a shows the transistor level simulation result for a 32×32 array with test image. Figure 7.20b shows the row-wise and column-wise summed currents (I_i and I_j), the outputs of row and column WTAs ($A_{1...V}$ and $A_{1...H}$), and V_{STOP}. As the shrink operation is started at 55 μs, I_i and I_j are started to be decreased continuously, as shown in Figure 7.20b. The currents that correspond to the smaller circle located at (21, 22) are completely reduced to zero at 60 μs. This means that the smaller circle disappears in the image. The logic output of the WTA changes to low as time elapses, and only the winning row and column are kept at a high logic. When the current of the 11th row and 9th column that corresponds to the center of the bigger circle is lower than I_{STOP}, V_{STOP} is changed to High, causing the winning address to be latched at 67 μs. The operation time of shrinking is about 12 μs, and it is decided automatically by the stop criterion circuit.

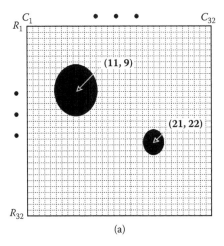

(a)

FIGURE 7.20 Transistor-level simulation results. (a) Input image for simulation (Copyright © Kluwer 2005.)   *(continued)*

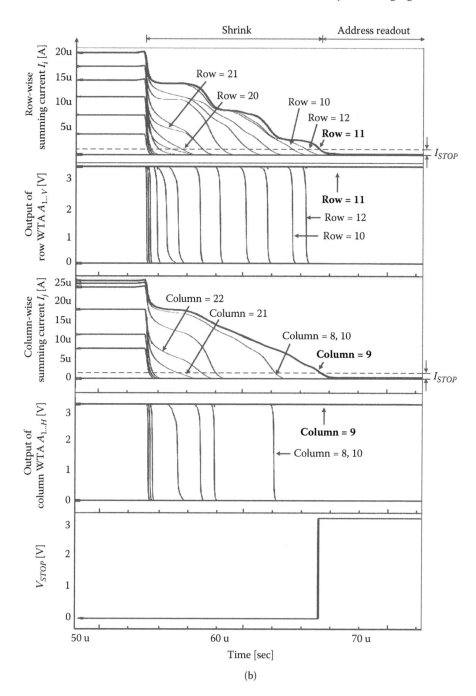

FIGURE 7.20 (CONTINUED) Transistor-level simulation results. (b) Simulated wave-forms during the shrink operation (Copyright © Kluwer 2005.)

The proposed single-chip eye tracker was fabricated with a 0.35 μm CMOS process. Figure 7.21 shows the microphotograph of the fabricated chip. The fabricated pixel array is 32×32, and one pixel occupies 50×50 μm^2. The photodiode is realized with an n$^+$/p$^-$ substrate junction whose area is 10×10 μm^2.

Figure 7.22 shows the testing setup for the error performance measurement with the fabricated prototype chip. The eye image captured under the infrared illumination is projected on the prototype chip through the beam projector and microscope. The light intensity of the image is decreased to a realistic intensity level by optical attenuation filters. The projected image on the chip is monitored by a digital camera instantaneously. The image on the pixel array is aligned precisely with the image displayed on the computer.

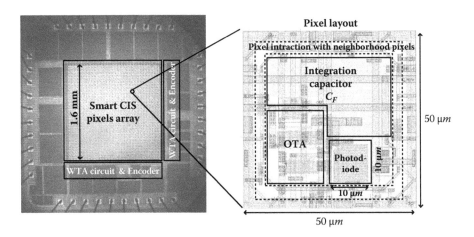

FIGURE 7.21 Chip microphotograph of the fabricated eye tracker and the layout of the smart pixel. (Copyright © Kluwer 2005.)

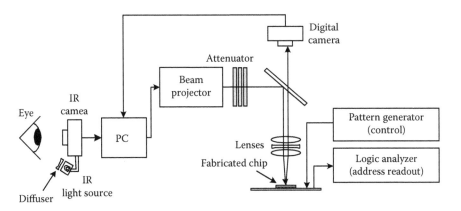

FIGURE 7.22 Testing environment for error measurement with the fabricated chip. (Copyright © Kluwer 2005.)

Figure 7.23 shows the microphotograph of the projected test images on the fabricated chip. The measurement results that were read out from the fabricated chip are marked as × and the expected results from the simulation are marked as □. Figure 7.24 shows the measured error map of the fabricated eye tracker. The error of the fabricated eye tracker was tested with test images that are similar to the test image shown in Figure 7.20(a). The center of the black circle is changed pixel-by-pixel and the output address from the chip is compared with the ideal one. The test result shows that the error is within ± 1 pixel. Error distribution in the upper-right side of the array is slightly higher than that of lower-left side of the array due to pro-cess variation. The frame rate is 125 frame-per-second (fps) and the power consumption is 260 mW with 3.3 V power supply. The specification of the prototype eye tracker is summarized in Table 7.3.

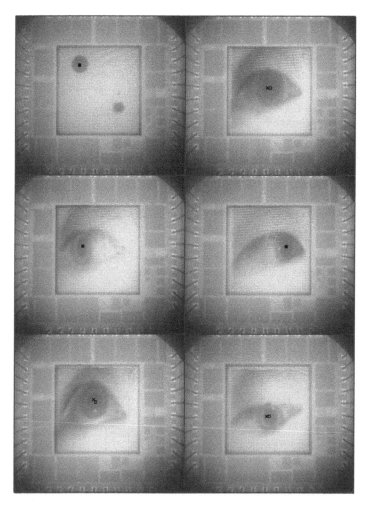

FIGURE 7.23 Image on the prototype chip and measurement results. (Copyright © Kluwer 2005.)

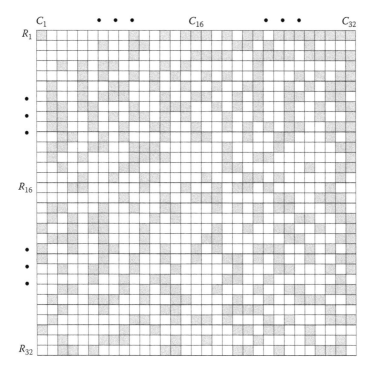

FIGURE 7.24   The measured error map of the fabricated eye tracker. (Copyright © Kluwer 2005.)

TABLE 7.3
Performance Summary of the First Version of the Eye Tracker

Process	0.35 µm CMOS 2-poly 4-metal
Power supply	3.3 V
Power consumption	260 mW
Chip size	1.95×1.95 mm^2
Smart CIS cell array	32×32
Smart CIS cell size	50×50 µm^2
Photodiode size	10×10 µm^2
Quantum efficiency (QE)	0.19 @ 1000 nm
PD signal range	1.9 V
Frame rate	125 fps
Integration time	1 ms
Error	± 1 pixel

7.3 SECOND VERSION OF THE EYE TRACKER

This section proposes the second version of the single-chip eye tracker that eliminates the glint effect and generates the digital address for the center of the pupil. Section 7.3.1 proposes an eye-tracking algorithm and architecture that can eliminate the effect of glints. Section 7.3.2 describes circuit implementation. Section 7.3.3 presents the simulation and experimental results. Finally, Section 7.3.4 provides the conclusion.

7.3.1 PROPOSED EYE-TRACKING ALGORITHM AND ARCHITECTURE

Figure 7.25 describes the algorithmic processes of eye tracking. The shrink operation is the process where each white pixel gives feedback to surrounding pixels that

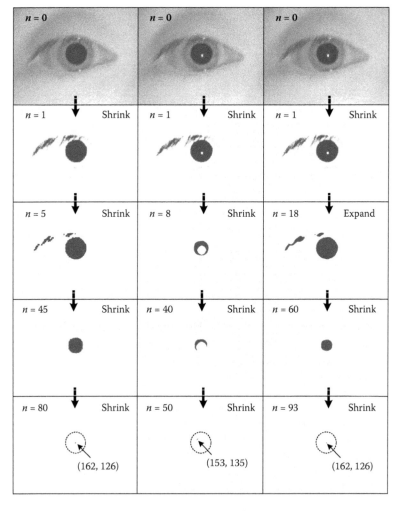

FIGURE 7.25 Algorithmic processes of eye tracking. Image size is 320×240, n is the iteration number, and the dashed circle indicates the shape of the pupil. (Copyright © IEEE 2009.)

forces them to be white, whereas the black pixel gives negligible feedback. As the shrink operation is performed repeatedly, the boundary's black pixels are continuously changed to white, and the size of the black region shrinks, eliminating the small black regions, such as iris, eyelashes, eyelid, and various shadows, as shown in Figure 7.25a. This process is repeated until only a few pixels remain. Then, the location of the remaining pixels represents the center of the pupil.

Although the shrink operation is an effective way to eliminate other objects and find the center of the pupil, the glint generated by the corneal reflection of the light source is inevitable, as shown in Figure 7.25b. Since this glint is expanded during the shrink operation and causes a wrong center address (Figure 7.25b), the white hole should be filled with black pixels to eliminate the effect of the glint. The hole filling can be performed by the *expand* operation, as shown in Figure 7.25c. During the expand operation, the black pixels force their neighbor pixels to be black, and this interaction causes the boundary's white pixels to change to black continuously. After a certain number of iterations of the shrink operation to eliminate noisy black objects, the expand operation is executed to fill the hole caused by the glint. Once the hole is filled with black pixels, the shrink operation resumes and continues until the last pixel remains. Although the number of iterations for expansion should be decided imperically, the introduction of the expand operation provides accurate center location.

The proposed eye tracker is composed of a smart pixel array, WTA circuits, a stop criterion circuit, latches, and address encoders, as shown in Figure 7.26. The smart pixel array captures the image and performs the shrink and expand operations. The number of black pixels is counted vertically and horizontally by summing the currents from pixels in each row or column, respectively. The WTA circuit not only finds the location of the largest current out of the column-wise summed current, I_i^V, or row-wise summed current, I_j^H, but also copies the winning current value, I_w^V or I_w^H. The stop criterion circuit generates the *latch* signal when any of I_w^V or I_w^H is smaller than the reference current, I_{STOP}, that corresponds to two black pixels. The WTA circuit generates high logic only for the winning column or row, while all other logic outputs (A_i or A_j) are low. Each latch stores the corresponding WTA logic output when the *latch* signal is generated by the stop criterion circuit. Then, the winning addresses are encoded and read out.

7.3.2   CIRCUITS IMPLEMENTATION

The smart CIS pixel array performs the shrink and expand operations with continuous-time interpixel feedback. Figure 7.27a presents the functional block diagram of the proposed smart pixel whose dynamics can be expressed as

$$\begin{cases} y_{ij} = f(x_{ij}) \\ n_{ij} = \sum_{k=-1}^{1}\sum_{l=-1}^{1} \varepsilon_{kl} y_{(i+k,j+l)} \\ C_{PD}\dfrac{dx_{ij}}{dt} = h_{s,e}(n_{ij}) \end{cases} \tag{7.6}$$

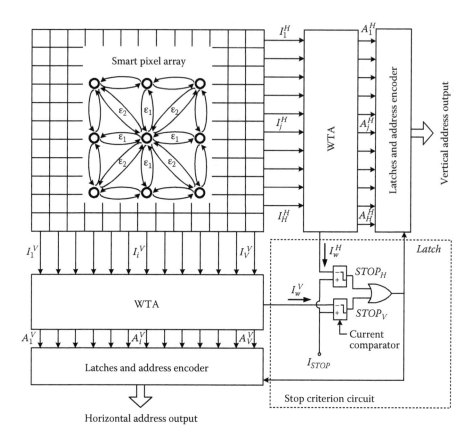

FIGURE 7.26 Block diagram of the proposed eye tracker. (Copyright © IEEE 2009.)

where C_{PD} is the parasitic capacitance of the PD and x_{ij} represents the state voltage whose initial value is decided from the captured image. Here, x_{ij} is lower for the brighter light. $f(x_{ij})$ is an inverted saturation function that discriminates the black pixel from the white pixel. n_{ij} is the weighted sum of the feedback from the neighboring pixels, as shown in Figure 7.28. Here, the weights for diagonal neighbors should be smaller than those for the vertical and horizontal neighbors to guarantee the isotropic shrink and expansion. $h_s(n_{ij})$ and $h_e(n_{ij})$ are nonlinear transconductance functions to generate feedback currents, I_{FB}, respectively, for the shrink and expand operations.

Each pixel counts the number of neighboring white pixels by calculating n_{ij}. When the control signal is set to perform the shrink operation, I_{FB} is generated based on $h_s(n_{ij})$, as shown in Figure 7.25a. If more than three white pixels exist in the neighborhood, then the pixel is considered to be located at the boundary of the black pixels. The transconductor takes out current from the C_{PD}, causing the x_{ij} to be lowered, which corresponds to it becoming a white pixel. On the contrary, when the control signal is set to perform the expand operation, I_{FB} is generated based on $h_e(n_{ij})$. If less than five white pixels exist in the neighborhood, then the pixel is considered to be located at the

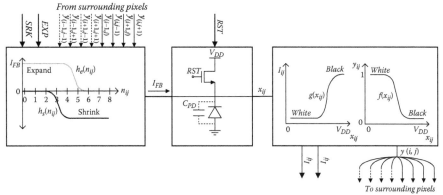

(a) Functional diagram of the proposed smart pixel.

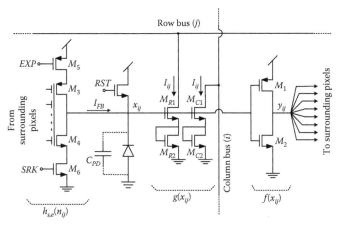

(b) Schematic diagram of the smart CIS pixel.

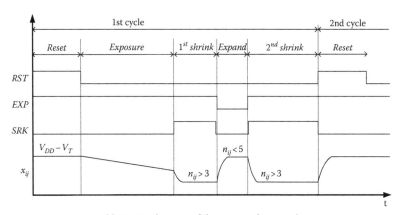

(c) Timing diagram of the proposed eye tracker.

FIGURE 7.27 Functional, schematic, and timing diagram of the smart CIS pixel. (Copyright © IEEE 2009.)

boundary of the white pixels. The transconductor supplies current to the C_{PD}, causing the x_{ij} to be increased, which corresponds to it becoming a black pixel.

Each pixel generates I_{ij} when the pixel is considered a black pixel by the saturation function $g(x_{ij})$, as shown in Figure 7.25a. The currents from pixels are summed column-wise and row-wise using the column-bus line and row-bus line, respectively, as shown in Figure 7.24. The summed currents, I_i^V and I_j^H, represent how many black pixels are in each column and row, respectively.

The schematic and timing diagram of the proposed smart pixel are shown in Figure 7.25b and c. First, the PD capacitor, C_{PD}, is precharged to $(V_{DD}-V_T)$ during the reset phase by turning on the RST switch. Once the RST is off, light-generated electrons are accumulated in C_{PD}, lowering the potential x_{ij} during the exposure phase. The transistors M_1 and M_2 form the nonlinear function, $f(x_{ij})$, whose output voltage is distributed to surrounding pixels. $M_{3,4,5,6}$ form the nonlinear transconductors, $h_s(n_{ij})$ and $h_e(n_{ij})$. It works as a current source that realizes $h_e(n_{ij})$ when M_5 is on and M_6 is off. On the contrary, it works as a current sink that realizes $h_s(n_{ij})$ when M_5 is off and M_6 is on. Both M_5 and M_6 are off and no interpixel interaction occurs during the exposure phase. $M_{R1,2}$ or $M_{C1,2}$ forms a nonlinear transconductor $g(x_{ij})$. Two stacked transistors are turned on and generate output current, I_{ij}, only when x_{ij} exceeds $2V_T$, which corresponds to the black pixel.

The weighted sum of the feedback from the neighboring pixels is obtained by a floating gate associated with M_3 and M_4. Figure 7.28 shows the schematic and layout of the weighted-sum circuit using floating gate capacitors. The floating gate voltage, v_{fg}, is obtained as follows:

$$v_{fg} = \sum_{k=-1}^{1} \sum_{l=-1}^{1} \left(\frac{C_{kl}}{C_{tot}} \cdot y_{(i+k,j+l)} \right) \tag{7.7}$$

where C_{kl} is the floating gate capacitance and C_{tot} is the sum of the floating gate capacitances and the parasitic gate capacitances of M_3 and M_4. Feedback coefficients,

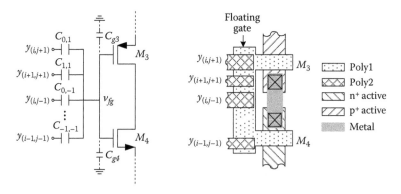

FIGURE 7.28 Schematic and layout of the weighted-sum circuit using floating gate capacitors. (Copyright © IEEE 2009.)

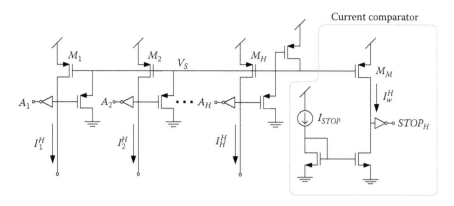

FIGURE 7.29 Circuit diagram of the WTA and current comparator. (Copyright © IEEE 2009.)

$\varepsilon_{k,l}$, in (7.6) are decided by C_{kl}, which is defined by the electrode area overlaid on the floating gate, as shown in Figure 7.26b.

Figure 7.29 presents the schematic diagram of the current mode WTA [22–26] circuit and the current comparator. The input currents of WTA are column-wise or row-wise summed currents, I_i^V or I_j^H. It is decided to turn on only the common source voltage, V_S, the winning transistor that has the highest input current out of $M_{1...H}$, and all others are turned off. Then the highest input current is copied to the current comparator. Since only the winning transistor is turned on and allows current flow, the logic output ($A_{1...H}$) is high only for the winner, and all others are low. Each logic output, A_i, is connected to its corresponding latch. The current comparator compares the winning current, I_w^H, with the reference current, I_{STOP}, and generates high logic output, $STOP_H$, when I_w^H is smaller than I_{STOP}.

An identical WTA circuit is used for the horizontal and vertical WTAs. The stop criterion circuit generates a high logic *latch* signal when either the horizontal or vertical current comparator generates a high logic $STOP_{H,V}$ signal, as shown in Figure 7.24. The *latch* signal generated by the stop criterion circuit makes the latches store the winning locations from the WTA circuits. The wining addresses are read out through the address encoder, which is implemented with the counter and the latch array. The counter counts the number of clocks until the shifted latch array output is high. Then, the counter contents represent the winner address.

7.3.3 Simulation and Experimental Results

Figure 7.30 shows the transistor level simulation results for a 50×40 pixel array. Figure 7.30a shows the images taken at $T_{0,1,\cdots,5}$, and Figure 7.30b shows the waveforms. After 1 ms of exposure time, the first shrink operation is started at T_0. and are continuously decreased because the small black regions around the pupil caused by the iris, eyelashes, eyelids, and various shadows are shrunk and disappear. The size of the glint is slightly increased until the expand operation is started at T_2. As the black region is expanded during the expand phase, and at (31, 23), that is, the

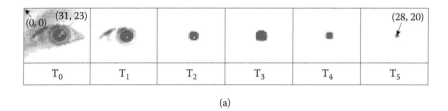

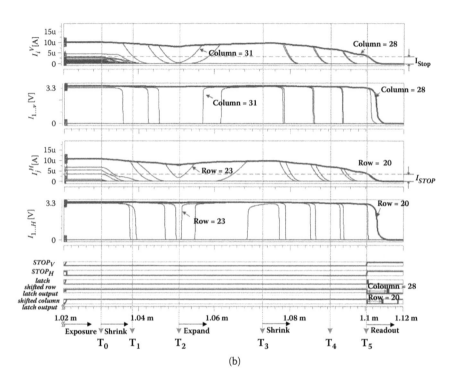

FIGURE 7.30 Transistor level simulation results. (a) Image transitions extracted from transistor simulation results: input eye image has a pupil center at (28, 20) and a glint near (31, 23). (b) Simulated waveforms for one frame. (Copyright © IEEE 2009.)

address of the glint, are increased. The glint is eliminated completely at T_3. The second shrink operation resumes at T_3 and the pupil shrinks again. The currents that correspond to the center of the pupil at (28, 20) remain the winner and reach to I_{STOP} at T_5, while all other currents become zero in advance. The *latch* signal is changed to High, causing the winning address to be latched.

The logic outputs of the WTA, A_i or A_j, change to low as time elapses, and only the winning row and column pixels are kept at high logic. The *shifted column latch output* and the *shifted row latch output* show the pulses that correspond to the winning location. The operation time to find the center of the pupil is about 100 μs.

The proposed single-chip eye tracker was fabricated with a 0.35 μm CMOS process. Figure 7.31 shows the microphotograph of the fabricated chip. The fabricated

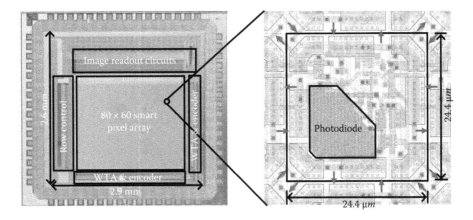

FIGURE 7.31 Chip microphotograph of the fabricated eye tracker and the layout of the smart pixel. (Copyright © IEEE 2009.)

chip has an array of 80×60 pixels with 24.4 μm pixel pitch, and the core area is 2.9×2.6 mm^2. The photodiode is realized with an n$^+$/p$^-$ substrate junction whose fill factor is about 20%. The image readout circuit that is composed of a 3-TR pixel circuit and the column-parallel single-slope ADC, including correlated double sampling (CDS) [27–29], are implemented as well for the testing purpose.

Figure 7.32 shows the testing setup for the performance measurement with the fabricated prototype chip. The fabricated eye tracker is mounted in an HMD. Infrared (IR) illumination and an IR filter are also assembled. The user gazes at a point on the screen of the HMD, and the eye image is captured by the fabricated chip through a half mirror of the HMD. The captured image is sent to the PC for testing purposes. At the same time, the eye-tracking output address from the fabricated chip is monitored by the logic analyzer.

Figure 7.33 shows the sample images taken from the fabricated chip and the detected center of the pupils. Besides the detected center output address from the fabricated sensor, the center of the pupil is also obtained by the software computation. Results of the software computation are marked as □, and the output addresses obtained from the fabricated chip are marked as ×. The error is defined

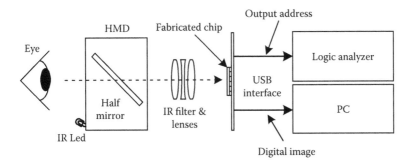

FIGURE 7.32 Testing environments with the fabricated chip.

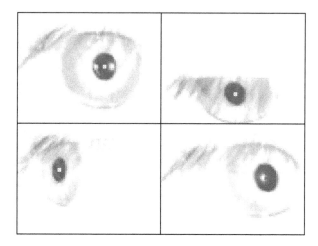

FIGURE 7.33   Sample image and the center location from the fabricated sensor. (Copyright © IEEE 2009.)

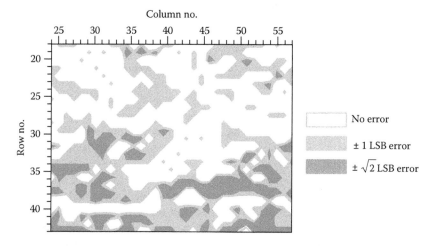

FIGURE 7.34   The measured error map from the fabricated eye tracker and the software computation. (Copyright © IEEE 2009.)

by the difference between the software computation result and the sensor output. Figure 7.34 shows the measured error map of the fabricated eye tracker that is obtained by gazing at a target at various locations on the screen. The test result shows that the error is within pixel pitch. The error in the lower region of the array is slightly higher than that in the other regions of the array because the pupil is slightly covered by the eyelash when the eye gazes toward the lower direction. The eye-tracking time is 200 μs excluding the exposure time, and the power consumption is 100 mW with a 3.3 V power supply. The specification of the prototype eye tracker is summarized in Table 7.4.

TABLE 7.4
Performance Summary of the Second Version of the Eye Tracker

Process	0.35 μm CMOS 2-poly 3-metal
Power supply	3.3 V
Power consumption	100 mW
Chip size	2.9×2.6 mm^2
Array size	80×60
Pixel size	24.4×24.4 μm^2
Photodetector	n$^+$/p$^-$ substrate
Fill factor	20%
Conversion gain	1.6 μV/e$^-$
Full well capacity	1.19×10^6 e$^-$
Sensitivity @ 950 nm	5.6 V/s $\cdot$ (μW/cm^2)
Sampling rate (tracking time)	5000 fps (200 μs)
Error	Less than $\pm\sqrt{2}$ pixel pitch

7.4 THIRD VERSION OF THE EYE TRACKER

This section proposes the third version of the single-chip eye tracker that has a 512×256 improved smart pixel array. Section 7.4.1 describes circuit implementation. Section 7.4.2 presents the simulation and experimental results. Finally, Section 7.4.3 provides the conclusion.

7.4.1 PROPOSED EYE-TRACKING ARCHITECTURE

The third version of the eye tracker has the same structure of the second version of the eye tracker, as shown in Figure 7.24. The eye tracker is composed of a smart pixel array, WTA circuits, a stop criterion circuit, latches, and address encoders. The smart pixel array captures the image and performs the shrink and expand operations. The number of black pixels is counted vertically and horizontally by summing the currents from pixels in each row or each column, respectively. The WTA circuit not only finds the location of the largest current out of the column-wise summed current, I_i^V, or row-wise summed current, I_j^H, but also copies the winning current value, I_w^V or I_w^H. The stop criterion circuit generates the *latch* signal when any of I_w^V or I_w^H is smaller than the reference current, I_{STOP}, that corresponds to two black pixels. The WTA circuit generates high logic only for the winning column or row, while all other logic outputs (A_i or A_j) are low. Each latch stores the corresponding WTA logic output when the *latch* signal is generated by the stop criterion circuit. Then, the winning addresses are encoded and read out.

7.4.2 CIRCUIT IMPLEMENTATION

The smart CIS pixel array performs the shrink and expand operations with continuous-time interpixel feedback. Figure 7.35a presents the functional block diagram of the proposed smart pixel whose dynamics can be expressed as

$$\begin{cases} y_{ij} = f(x_{ij}) \\ n_{ij} = \displaystyle\sum_{k=-1}^{1}\sum_{l=-1}^{1} \varepsilon_{kl}\, y_{(i+k,j+l)} + \alpha \cdot OPC \\ C_{PD}\dfrac{dx_{ij}}{dt} = h_{s,e}(n_{ij}) \end{cases} \qquad (7.8)$$

where C_{PD} is the parasitic capacitance of the PD and x_{ij} represents the state voltage whose initial value is decided from the captured image. Here, x_{ij} is lower for brighter light. $f(x_{ij})$ is an inverted saturation function that discriminates the black pixel from the white pixel. The n_{ij} is the weighted sum of the feedback from the neighboring pixels. Here, the weights for diagonal neighbors should be smaller than those for vertical

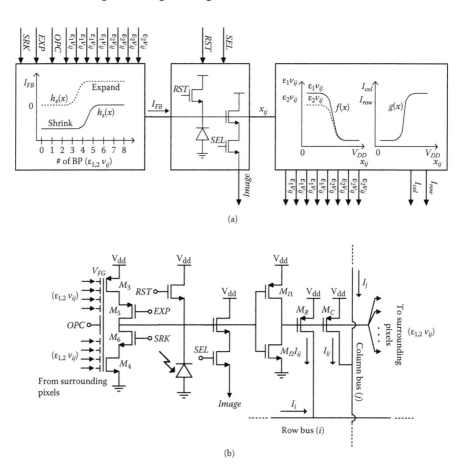

(a)

(b)

FIGURE 7.35 Functional and schematic diagrams of the proposed smart pixel: (a) functional diagram and (b) schematic diagram.

and horizontal neighbors to guarantee the isotropic shrink and expansion. $h_s(n_{ij})$ and $h_e(n_{ij})$ are nonlinear transconductance functions to generate feedback currents, I_{FB}, respectively, for the shrink and expand operations.

Each pixel counts the number of neighboring white pixels by calculating n_{ij}. When the control signal is set to perform the shrink operation, I_{FB} is generated based on $h_s(n_{ij})$, as shown in Figure 7.35a. If more than three white pixels exist in the neighborhood, then the pixel is considered to be located at the boundary of the black pixels. The transconductor takes out current from the C_{PD}, causing n_{ij} to be lowered, which corresponds to becoming a white pixel. On the contrary, when the control signal is set to perform the expand operation, I_{FB} is generated based on $h_e(n_{ij})$. If fewer than five white pixels exist in the neighborhood, then the pixel is considered to be located at the boundary of the white pixels. The transconductor supplies current to the C_{PD}, causing the x_{ij} to increase, which corresponds to becoming a black pixel.

Each pixel generates I_{ij} when the pixel is considered a black pixel by the saturation function $g(x_{ij})$, as shown in Figure 7.35a. The currents from pixels are summed column-wise and row-wise using column-bus line and row-bus line, respectively, as shown in Figure 7.24. The summed currents, I_i^V and I_j^H, represent how many black pixels are in each column and row, respectively.

The schematic of the proposed smart pixel is shown in Figure 7.35b. First, the PD capacitor, C_{PD}, is precharged to $(V_{DD}-V_T)$ during the reset phase by turning on the RST switch. Once the RST is off, light-generated electrons are accumulated in C_{PD}, lowering the potential x_{ij} during the exposure phase. The transistors M_{I1} and M_{I2} form the nonlinear function, $f(x_{ij})$, whose output voltage is distributed to surrounding pixels. $M_{3,4,5,6}$ form the nonlinear transconductors, $h_s(n_{ij})$ and $h_e(n_{ij})$. It works as a current source that realizes $h_e(n_{ij})$ when M_5 is on and M_6 is off. On the contrary, it works as a current sink that realizes $h_s(n_{ij})$ when M_5 is off and M_6 is on. Both M_5 and M_6 are off and no interpixel interaction occurs during the exposure phase. M_R or M_C forms a nonlinear transconductor $g(x_{ij})$. Two stacked transistors are turned on and generate output current, I_{ij}, only when the pixel is black.

If I_{FB} is large, the shrink and expand operations perform too fast because the parasitic capacitance of the photodiode is very small. I_{FB} must be controlled as a small value to latch the last winner location. Therefore, M_5 and M_6 of the second version of the smart pixel in Figure 7.25 are operated in the subthreshold region by the EXP and SRK. The EXP and SRK are about $(V_{DD}-V_T)$ and VT values, respectively. However, the third version of the eye tracker does not need these analog controls. The OPC and operation capacitor, C_{OPC}, enable M_5 and M_6 transistors to operate in the subthreshold region.

The weighted sum of the feedback from the neighboring pixels is obtained by a floating gate associated with M_3 and M_4. The floating gate voltage, v_{fg}, is obtained as follows:

$$v_{fg} = \sum_{k=-1}^{1}\sum_{l=-1}^{1}\left(\frac{C_{kl}}{C_{tot}}\cdot y_{(i+k,j+l)}\right) + \frac{C_{OPC}}{C_{tot}}\cdot OPC \qquad (7.9)$$

where C_{kl} is the floating gate capacitance, C_{OPC} is the operation capacitance, and C_{tot} is the sum of the floating gate capacitances and the parasitic gate capacitances of M_3 and M_4.

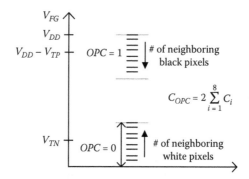

FIGURE 7.36 Potential of the floating gate.

Feedback coefficients, $\varepsilon_{k,l}$, α, in (7.8) are decided by C_{kl}, which is defined by the electrode area overlaid on the floating gate. The voltage of the floating gate is shown in Figure 7.36.

Figure 7.37 presents the schematic diagram of the current mode WTA [23–26] circuit and the current comparator. The input currents of WTA are column-wise or row-wise summed currents, I_i^V or I_j^H. It is decided to turn on only the common source voltage, V_S, the winning transistor that has the highest input current out of $M_{1...H}$, and all others are turned off. Then the highest input current is copied to the current comparator. Since only the winning transistor is turned on and allows current flow, the logic output ($A_{1...H}$) is high only for the winner, and all others are low. Each logic output, A_i, is connected to the corresponding latch. The current comparator compares the winning current, I_w^H, with the reference current, I_{STOP}, and generates high logic output, $STOP_H$, when I_w^H is smaller than I_{STOP}.

An identical WTA circuit is used for the horizontal and vertical WTAs. The stop criterion circuit generates a high logic *latch* signal when either the horizontal or vertical current comparator generates a high logic $STOP_{H,V}$ signal, as shown in Figure 7.24. The *latch* signal generated by the stop criterion circuit makes the latches store the winning locations from the WTA circuits. The winning addresses are read out though the address encoder, which is implemented with the counter and the latch array. The counter counts the number of clocks until the shifted latch array output is high. Then, the counter contents represent the winner address.

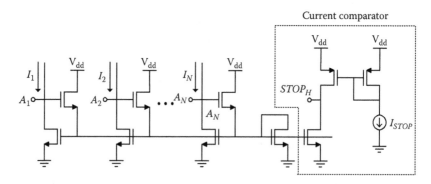

FIGURE 7.37 Circuit diagram of the WTA and current comparator.

7.4.3 EXPERIMENTAL RESULTS

The proposed single-chip eye tracker was fabricated with a 0.35 μm CMOS process. Figure 7.38 shows the microphotograph of the fabricated chip and the core area is 9.3×2.6 mm². The fabricated pixel array is 512×256, and one pixel occupies 15.8×18.0 μm². The photodiode is realized with an n⁺/p⁻ substrate junction whose fill factor is about 31%. The fabricated third version of the eye tracker is under testing.

Figure 7.39 shows the expected testing setup for the performance measurement with the fabricated prototype chip and HMD. The user looks at the screen of the HMD and

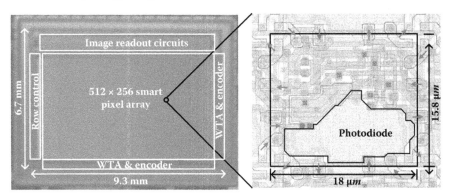

FIGURE 7.38 Chip microphotograph of the fabricated eye tracker and layout of the smart pixel. Pixel interaction lines with neighborhood pixels are indicated by the arrows.

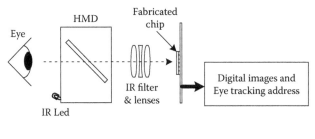

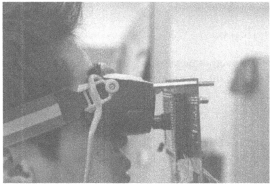

FIGURE 7.39 Expected testing environment for error measurement with the fabricated chip.

FIGURE 7.40 (See color insert.) The expected measurement results of the fabricated chip. Circles on the image indicate visual fixations, the size of the circle implies the fixation time, and arrows show the saccades.

the eye image under infrared illumination is captured through a half mirror of the HMD, IR filter, and lenses. The captured image is converted to the digital image in image sensor mode, and the eye-tracking address is output in the eye-tracking mode.

The expected measurement results of the fabricated chip linked with an HMD are presented in Figure 7.40. A landscape is showed to a user in HMD, and then the eye movement is tracked for 3 s. The expected results are accomplished with eye-gazing location and time, which are output from the eye tracker every 2 ms. Circles on the image indicate the visual fixations, and the size of circles implies the fixation time. Arrows show the saccadic eye movements.

7.5 CONCLUSION

This chapter introduced three versions of high-speed single-chip eye trackers using smart CIS pixels that do not require any additional peripherals. The proposed sensors detect the center of the pupil using the shrink/expand operations and WTA that were realized with compact analog circuits. The shrink operation effectively removes noisy objects, whereas the expand operation effectively removes the glints, providing the center of the pupil.

The first version of the eye tracker was designed and fabricated with a 0.35 μm CMOS process having a 32×32 smart CIS pixel array, and the core size was 3.8 mm². As the frame rate of the proposed eye tracker is 125 fps and the conventional eye tracker is commonly 30 fps, the prototype chip test results demonstrated that the proposed system can generate an address indicating on what the user gazes at a higher speed than the conventional one. Measurement results show ±1 pixel error, and the power consumption is 260 mW. The pixel size and power consumption of the first version of the eye tracker are still large, and there is a possibility to improve the pixel size.

The second version of the eye tracker was designed and fabricated with a 0.35 μm CMOS process having an 80×60 smart CIS pixel array, and the core size was 7.5 mm^2. The second version of the eye tracker eliminates the glint effect caused by the illumination light. The test results show pixel error at the rate of 5000 fps. The power consumption is 100 mW with a 3.3 V supply voltage. Even though the smart pixel of the second version performs more complex functions, the pixel area of the second version of the eye tracker is decreased to 23.8%, compared to the pixel area of the first version of the eye tracker. The power consumption is reduced to 38% of the power consumption of the first version. However, the pixel array is still small, and there is a possibility to improve the resolution.

The third version of the eye tracker was designed and fabricated with a 0.35 μm CMOS process having a 512×256 smart CIS pixel array, and the core size was 62.3 mm^2. The third version of the eye tracker has a higher spectification than that of the commercial eye tracker. The expected test results show a pixel error at the rate of 1000 fps. The power consumption is 200 mW with a 3.3 V supply voltage. The pixel area of the third version of the eye tracker is 47.7% of the second version and 11% of the first version.

The introduced eye trackers are intended to be used as a pointing device in conjunction with an HMD and an input device for the handicapped. Therefore, the handicapped can give instructions to the computer just by gazing at the location, and more interactive video games can be accomplished by HMD, including the eye tracker.

REFERENCES

1. T. Miyoshi and A. Murata, Input device using eye tracker in human-computer interaction, in *Proceedings of the 10th IEEE International Workshop on Robot and Human Interactive Communication*, Bordeaux, Paris, France, September 2001, pp. 18–21.
2. G. Beach, C. J. Cohen, J. Braun, and G. Moody, Eye tracker system for use with head mounted displays, in *IEEE International Conference on Systems, Man, and Cybernetics*, Tokyo, October 1999, vol. 5, pp. 4348–4352.
3. K. Iwamoto, S. Katsumata, and K. Tanie, An eye movement tracking type head mounted display for virtual reality system: Evaluation experiments of a prototype system, in *IEEE International Conference on Systems, Man, and Cybernetics*, San Antonio, Texas, October 1994, vol. 1, pp. 13–18.
4. Z. Zhiwei, J. Qiang, K. Fujimura, and L. Kuangchih, Combining Kalman filtering and mean shift for real time eye tracking under active IR illumination, in *Proceedings of the 16th International Conference on Pattern Recognition*, 2002, vol. 4, pp. 318–321.
5. T. Oya, H. Hashimoto, and F. Harashima, Active eye sensing system-predictive filtering for visual tracking, in *Proceedings of the 16th International Conference on Industrial Electronics, Control, and Instrumentation*, November 1993, vol. 3, pp. 1718–1723.
6. A. Graupner, J. Schreiter, S. Getzlaff, and R. Schüffny, CMOS image sensor with mixed-signal processor array, *IEEE J. Solid-State Circuits*, 38, 948–957, 2003.

7. Y. Muramatsu, S. Kurosawa, M. Furumiya, H. Ohkubo, and Y. Nakashiba, A signal-processing CMOS image sensor using a simple analog operation, *IEEE J. Solid-State Circuits*, 38, 101–106, 2003.

8. S. Espejo, A. Rodríguez-Vázquez, R. Domínguez-Castro, J. L. Huertas, and E. Sánchez-Sinencio, Smart-pixel cellular neural networks in analog current-mode CMOS technology, *IEEE J. Solid-State Circuits*, 29, 895–905, 1994.

9. M. Schwarz, R. Hauschild, B. J. Hosticka, J. Huppertz, T. Kneip, S. Kolnsberg, L. Ewe, and H. K. Trieu, Single-chip CMOS image sensors for a retina implant system, *IEEE Trans. Circuits Syst. II*, 46, 870–877, 1999.

10. L. O. Chua and L. Yang, Cellular neural networks: Theory, *IEEE Trans. Circuits Syst. I*, 35, 1257–1272, 1988.

11. L. O. Chua and L. Yang, Cellular neural networks: Applications, *IEEE Trans. Circuits Syst. I*, 35, 1273–1290, 1988.

12. D. A. Robinson, The use of control system analysis in the neurophysiology of eye movements, *Ann. Rev. Neurosci.*, 4, 463–503, 1981.

13. IOTA AB, EyeTrace Systems, Sundsvall Business and Technology, http://www.iota.se

14. A. T. Duchowski, *Eye tracking methodology: Theory and practice*, Spring Verlag, New York, 2002.

15. Metrovision, France, http://www.metrovision.fr

16. Skalar Medical, Netherlands, http://www.skalar.nl

17. Microguide, USA, http://www.eyemove.com

18. Applied Science Group Company, Choosing an eye tracking system. http://http://www.a-s-l.com

19. J. Solhusvik et al., Recent experimental results from a CMOS active pixel image sensor with photodiode and photogate pixel, *Proc. SPIE*, 2950, 19–25, 1996.

20. S. Wolfram, *MATHEMATICA: A system for doing mathematics by computer*, Addison-Wesley, Reading, MA, 1988.

21. B. W. Char et al., *Maple V library reference manual*, Springer-Verlag, New York, 1991.

22. A. Fish, D. Turchin, and O. Yadid-Pecht, An APS with 2-D winner-take-all selection employing adaptive spatial filtering and false alarm reduction, *IEEE Trans. Electron Devices*, 50, 159–165, 2003.

23. T. Serrano-Gotarredona and B. Linares-Barranco, A high-precision current-mode WTA-MAX circuit with multichip capability, *IEEE J. Solid-State Circuits*, 33, 280–286, 1998.

24. I. E. Opris, Analog rank extractors, *IEEE Trans. Circuits Syst. I*, 44(12), 1114–1121, 1997.

25. J. Choi and B. J. Sheu, A high-precision VLSI winner-take-all circuit for self-organizing neural networks, *IEEE J. Solid-State Circuits*, 28(5), 576–584, 1993.

26. D. Kim, S. Lim, and G. Han, Single-chip eye tracker using smart CMOS image sensor pixels, *Analog Integr. Circuits Signal Process.*, 45, 131–141, 2005.

27. P. Lee, R. Gee, M. Guidash, T. Lee, and E. R. Fossum, An active pixel sensor fabricated using CMOS/CCD process technology, presented at *IEEE Workshop on CCD's and Advanced Image Sensors,* Dana Point, CA, April 20–22, 1995.

28. W. Yang, O. Kwon, J. Lee, G. Hwang, and S. Lee, An integrated 800×600 CMOS imaging system, *ISSCC Digest of Technical Papers*, February 1999, pp. 304–305.

29. T. Sugiki, S. Ohsawa, H. Miura, M. Sasaki, and N. Nakamura, A 60-mW 10-b CMOS image sensor with column-to-column FPN reduction, *ISSCC Digest of Technical Papers*, February 2000, pp. 108–109.

8 Laser Doppler Velocimetry Technology for Integration and Directional Discrimination

Koichi Maru and Yusaku Fujii

CONTENTS

8.1 INTRODUCTION

Laser Doppler velocimeters (LDVs) have been widely used to measure the velocity of a fluid flow or rigid object in various studies and industries since the introduction of the concept in 1964 [1]. Measurement by using differential LDV has the advantage of a contactless, small measuring volume, giving excellent spatial resolution and a linear response. However, conventional LDVs using bulk optical systems or fibers have large sizes and complex assemblies, and are often affected by environmental disturbances, such as vibrations, due to the large optical path length in the optical system. Hence, integrated LDVs with high precision and compact size have been in high demanded. Using a planar lightwave circuit (PLC) [2], in which several optical elements are arranged on a planar surface of a silica or silicon substrate, is a promising way to integrate an optical system in a small size. The PLC technology has been

189

widely used for optical passive devices deployed for optical communication systems because of its reliability and ability to be manufactured in large volumes.

Meanwhile, multidimensional velocity needs to be measured in many applications of fluid experiments, whereas many techniques for LDVs for measuring one-dimensional velocity have been developed [3]. LDVs using a polarization method [4–6], different wavelengths [7, 8], frequency shift discrimination [9–12], and a self-mixing technique based on dynamical perturbation of a laser [13] have been developed for measuring two or three velocity components. In typical conventional LDVs for measuring multidimensional velocity, a complicated optical configuration or a three-dimensional arrangement of optical components is needed. When we use wavelength discrimination, lasers with different wavelengths and color filters or color splitters are needed. In some LDVs based on frequency shift discrimination, frequency shifters such as acousto-optic modulators (AOMs) are needed to discriminate the direction of the velocity. Hence, the accuracy considerably depends on the performance of the frequency shifters to be used, and its measureable velocity range is limited by preshifting frequencies.

In this chapter, several types of integrated LDVs using PLC technology, especially a wavelength-insensitive LDV [14–17] and a multipoint LDV [18] using arrayed waveguide gratings (AWGs), are described. AWGs have been widely used and deployed as key filtering devices in commercial wavelength-division multiplexing (WDM) optical communication systems. In addition, techniques for two-dimensional velocity measurement using a simple optical configuration without multiple colors or any optical modulator [19, 20] are described.

8.2 INTEGRATED LDVS USING PLC TECHNOLOGY

8.2.1 ARRAYED WAVEGUIDE GRATING

The AWG is a planar device. The concept of the AWG was first proposed in 1988 by Smit [21]. Also in 1988, Dragone [22] reported the star coupler configuration that has been mainly used as an element of the AWG. Since the early 1990s, much attention has been paid to the realization of devices with superior performance, as well as their application to several functional optical circuits for WDM systems. Takahashi et al. [23, 24] reported the first devices operating in the long-wavelength window. Vellekoop and Smit [25] demonstrated the first devices operating in short wavelengths. Dragone extended the concept from $1 \times N$ to $N \times N$, that is, so-called WGRs [26, 27], which play an important role in the wavelength routing network. Since 1993, system experiments involving AWGs have been reported [28, 29].

Figure 8.1 illustrates the basic optical circuit of an AWG. The optical circuit consists of input waveguides, two slab waveguides (input and output slabs), a waveguide array, and output waveguides. The waveguide array is designed with waveguides having a constant waveguide length difference to adjacent waveguides. The two slabs have a function similar to that of lenses, and the waveguide array has a function similar to that of a grating. When the lightwave is launched into one of the input waveguides, it spreads out in the input slab and is coupled to the waveguides in the array. After passing through the array, each beam of lightwave interferes constructively

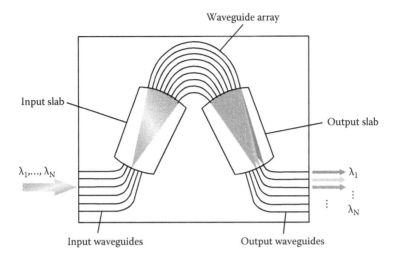

FIGURE 8.1 (See color insert.) Basic optical circuit of AWG.

or destructively according to the phase condition in the output slab. The constructively interfering lightwave focuses on one of the output waveguides according to its wavelength.

Silica-based PLC technology [2] has been used to fabricate waveguide devices, including AWGs. In a typical structure, the core buried by the cladding has a higher refractive index than that of the cladding by doping germanium, titanium, phosphorus, or boron into silica (SiO_2). The silica-based waveguides are formed on silica or silicon substrate by glass film deposition, including flame hydrolysis deposition (FHD) [30], radio frequency (RF) spattering [31], and chemical vapor deposition (CVD) [32], which is followed by the etching process. Passive devices using silica-based PLC have been widely used in optical communication systems because of their reliability, low insertion loss, ease of coupling to optical fibers, integration capacity, and ability to produce optical filters with high accuracy [33].

8.2.2 WAVELENGTH-INSENSITIVE LDV

For an integrated LDV, using a semiconductor laser is desirable as a light source for integration. However, typical inexpensive semiconductor lasers suffer from the problem of instability in lasing wavelength due mainly to the dependence on temperature. In conventional differential LDVs, the Doppler frequency shift at a monitoring point depends on the signal wavelength to be used. Hence, wavelength instability in semiconductor lasers causes measurement errors in the Doppler frequency shift. To reduce the measurement error due to wavelength change, differential LDVs with diffractive gratings have been reported [34–36]. In these LDVs, the dependence of the diffraction angle on wavelength is utilized for the wavelength-insensitive operation. However, this type of LDV needs assembly and alignment of a diffractive grating and other optical elements. To reduce the size and cost of devices, wavelength-insensitive LDVs consisting of integrated optical waveguides are desirable.

In typical differential LDV, Doppler frequency shift is sensitive to the signal wavelength of the input laser beam. The Doppler frequency shift F_D is expressed as

$$F_D = \frac{2v_\perp \sin \psi}{\lambda} \quad (8.1)$$

where ψ is the incident angle of the beam to the object, $v_\perp$ is the velocity of the object perpendicular to the bisector of the angles of the incident beams to the object, and λ is the wavelength. From this equation, if ψ appropriately changes depending on λ, the wavelength-insensitive operation can be expected. When F_D is to be wavelength insensitive around $\lambda = \lambda_0$, the derivative of F_D with respect to λ should be zero at $\lambda = \lambda_0$.

There are several methods to change the incident angle of the beam to the object ψ according to the wavelength change. One method to obtain the appropriate change of ψ is the LDV using AWGs. Figure 8.2 illustrates the optical circuit of the wavelength-insensitive LDV using AWGs [15, 17]. Input light is split with a 50:50 beam splitter. Each light is passing through each AWG, output with diffraction, and incident on the object. The beams are scattered on the object and detected by a photodetector (PD). Each AWG is used to diffract the beam whose diffraction angle is changed depending on the wavelength. Each AWG is also used to focus the beam to the object. In other words, the AWGs can function as a lens system and dispersive elements having a planar structure.

In the proposed structure, ψ is changed with the system using the AWGs. Hence, wavelength-insensitive operation is expected if the design of these AWGs is optimized. Provided that the upper AWG and the lower AWG have the same design parameters, the condition of the design parameters for wavelength-insensitive operation is given by $\tan\psi_0 = m\lambda_0/(n_s d)$, where n_s is the effective refractive index for the

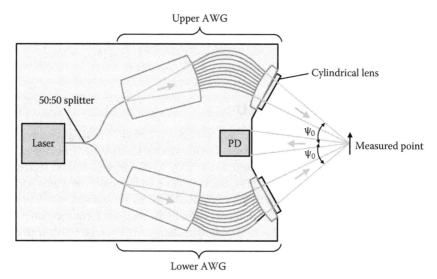

FIGURE 8.2 (See color insert.) Optical circuit of wavelength-insensitive LDV using AWGs. (From K. Maru and Y. Fujii, *Appl. Mechanics Mater.*, 103, 76–81, 2011.)

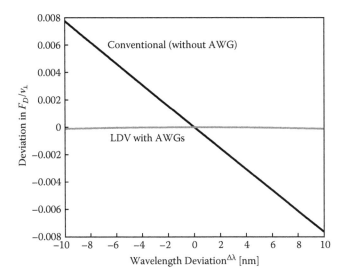

FIGURE 8.3 Deviation in $F_D/v_\perp$ for wavelength-insensitive LDV with AWGs as a function of wavelength deviation $\Delta\lambda = \lambda - \lambda_0$. (From K. Maru and Y. Fujii, *Appl. Mechanics Mater.*, 103, 76–81, 2011.)

slab waveguides in the AWGs, d is the interval of the waveguides in the array at the output side of the slab-to-array interface, and m is the grating order.

Figure 8.3 plots the deviation in $F_D/v_\perp$ determined as $[F_D/v_\perp - (F_D/v_\perp)|_{\lambda=\lambda_0}]/$ $(F_D/v_\perp)|_{\lambda=\lambda_0}$ as a function of the wavelength deviation $\Delta\lambda = \lambda - \lambda_0$. For comparison, an LDV without a wavelength-insensitive structure (i.e., without AWGs) is also shown in Figure 8.3. The absolute value of the deviation in $F_D/v_\perp$ can be significantly reduced to less than 1×10^{-4} by using AWGs, whereas the maximum deviation for the conventional structure without AWGs is 7×10^{-3} when the wavelength deviation $\Delta\lambda$ is within 9.0 nm.

8.2.3 MULTIPOINT LDV

In some cases such as fluid flow in narrow pipes, velocity distribution in depth direction should be measured. In this case, the velocities of different points in the depth direction should be simultaneously measured. For this purpose, several types of LDVs for multipoint measurement have been proposed [3, 37]. However, these multipoint LDVs consist of large bulk optical systems. Hence, an integrated type of multipoint LDVs has been highly demanded. Several integrated differential LDVs have been proposed [38, 39] as applications of integrated optical sensors, although these LDVs are used for single-point measurement.

Figure 8.4 illustrates the configuration of the proposed integrated multipoint differential LDV [18]. The LDV consists of laser diodes (LDs) with different lasing wavelengths as light sources, power splitters, two laser-side AWGs with the same design, cylindrical lenses, a detector-side AWG, and PDs. The beam from each laser is split with a beam splitter. Each spilt beam is incident on the input slab of each

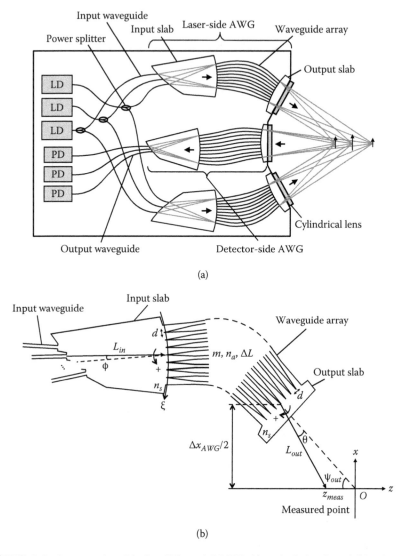

(a)

(b)

FIGURE 8.4 Integrated multipoint differential LDV: (a) optical circuit and (b) schematic diagram of laser-side AWG. (From K. Maru, K. Kobayashi, and Y. Fujii, *Opt. Express*, 18(1), 301–308, 2010.)

laser-side AWG, phase-shifted through its waveguide array, output with diffraction, and incident on the measured point. The beam is scattered on the object at the measured point, input to the detector-side AWG, diffracted according to its wavelength, and detected by one of the PDs. The cylindrical lenses are used to collimate the beams in the vertical direction. The beams with different wavelengths are focused on different focusing points by the laser-side AWGs, and detected by different PDs by the detector-side AWG. Hence, the velocities at multiple measured points can be simultaneously measured. The laser-side AWGs are also used to diffract the beam whose diffraction angle is changed depending on the wavelength in order to reduce the dependence of the measured Doppler shift on the wavelength. This change in the diffraction angle contributes to reducing wavelength sensitivity, as described in Section 8.2.2.

In order to investigate the design and characteristics of the circuit, the model illustrated in Figure 8.4b is derived. The simulation was performed by assuming use of silica-based materials as the optical circuit [18]. Figure 8.5 shows the relation between the relative position of the measured point in the depth direction, $z_{meas}/\Delta x_{AWG}$, and the input wavelength λ for various input angles φ. The lines for the constant θ are also plotted in Figure 8.5. Obviously, when the diffraction angle θ is not changed, the measured position $z_{meas}/\Delta x_{AWG}$ becomes a constant. If the beam is diffracted to the direction out of the central Brillouin zone [40], its power is significantly decreased and the beam becomes no longer the main beam. The input beam to the waveguide array also suffers from considerable loss when the beam direction is out of the central Brillouin zone. Hence, the input angle φ and diffraction angle θ for each input wavelength must be within the central Brillouin zone defined as the angle within $\pm\Delta\theta_B$. The condition of angles φ and θ within $\pm\Delta\theta_B$ is indicated as the area within the dotted line. The positions of measured points can be derived from this figure once a set of input angles φ and input wavelengths λ is determined. Table 8.1 shows an example of the parameters φ, λ, L_{in}, and z_{meas} for five-point velocity measurement. Here, we assume $\Delta x_{AWG} = 30$ mm. In this case, the measured points are arranged over the range of 25.77 mm.

The absolute value of deviation in $F_D/v_\perp$ can be reduced for the multipoint LDV. We have simulated the absolute value of deviation in $F_D/v_\perp$ for the parameters shown in Table 8.1 as the nominal input wavelengths and the input angles. The simulation

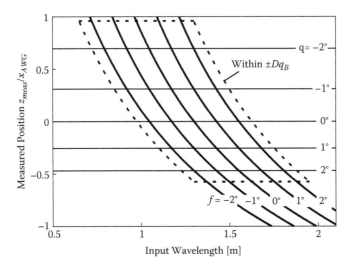

FIGURE 8.5 Relation between relative position of measured point $z_{meas}/\Delta x_{AWG}$ and input wavelength λ for various φ and θ. $m = 2$, $d = 10$ μm and $\psi_{out} = 10.17°$. The condition of angles φ and θ within $\pm\Delta\theta_B$ is indicated as the area within the dotted line. (From K. Maru, K. Kobayashi, and Y. Fujii, *Opt. Express*, 18(1), 301–308, 2010.)

TABLE 8.1

Example of Parameters φ, λ, L_{in}, and z_{meas}

Input Wavelength λ (μm)	Input Angle φ (degree)	Input–Array Distance L_{in} (mm)	Measured Position z_{meas} (mm)
1.20	0.7096	17.66	14.77
1.25	0.3548	17.93	6.80
1.30	0.0000	18.21	0.00
1.35	−0.3548	18.50	−5.87
1.40	−0.7096	18.80	−11.00

Source: K. Maru, K. Kobayashi, and Y. Fujii, *Opt. Express*, 18(1), 301–308, 2010.

result indicates that the absolute value of the deviation in $F_D/v_\perp$ for $\lambda = 1.3$ μm (i.e., the wavelength of the beam from the central input port) for the proposed structure can be significantly reduced to less than 10^{-4} of that for a conventional LDV, and that the deviation for $\lambda = 1.2$ and 1.4 μm (i.e., beams from marginal input ports) can also be reduced to less than 1/10 of that for a conventional LDV.

8.3 DIRECTIONAL DISCRIMINATION

Simultaneously measuring two components of velocity at a measured point has been required in many applications, such as fluid experiments. Several methods for measuring multidimensional velocity components have been developed [4–13]. In typical conventional LDVs for measuring two-dimensional velocity, a complicated optical configuration or a three-dimensional arrangement of optical components is needed. In some LDVs using frequency shift, frequency shifters such as AOMs are needed to discriminate the direction of velocity. In these LDVs, measurable velocity range is limited by preshifting frequencies, and the accuracy considerably depends on the performance of the optical modulators to be used. Several approaches have been reported for directional discrimination without optical modulators [39, 41], although these approaches are for one-dimensional velocity measurement. In this section, two methods for two-dimensional velocity measurement using a simple optical configuration without any optical modulator are reviewed.

8.3.1 TWO-DIMENSIONAL LDV USING POLARIZED BEAMS AND 90° PHASE SHIFT

Figure 8.6 illustrates the principle of the proposed LDV for two-dimensional velocity measurement using polarized beams and a 90° phase shift [20]. The beam output from a laser is split into a signal beam to be incident on the measured point and a reference beam by a 1 × 2 splitter. The signal beam is split again into two orthogonally polarized beams by the polarization beam splitter 1 (PBS1). The two beams are incident on the measured point with different angles, scattered on the object at the measured point, and split again by PBS2. After passing through PBS2, the beam is split into the scattered beam from port A and that from port B. Each scattered beam

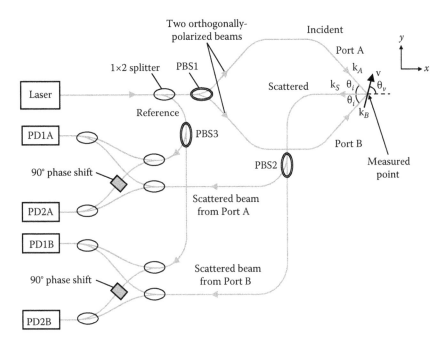

FIGURE 8.6 Principle of LDV for two-dimensional velocity measurement using polarized beams and 90° phase shift. (From K. Maru, L. Y. Hu, R. S. Lu, Y. Fujii, and P. P. Yupapin, *Optik*, 122(11), 974–977, 2011.)

is split into two beams, and each of the split beams is combined with one of the reference beams split by PBS3 and other 1×2 splitters. Here, before the reference beams are combined with each of the scattered beams, the phase of the reference beams is shifted so that the phase difference between the two reference beams becomes 90°. The power of each combined beams is detected by a PD as a beat signal produced by the interference of the scattered beam and the reference beam. For each scattered beam (from port A and from port B), two PDs and the 90° phase shift of the reference beam are used to detect the signs of the beat frequency and discriminate the direction of velocity.

The detected beat frequencies f_A at PD1A and PD2A and f_B at PD1B and PD2B are given by the wavenumber vectors of the incident beam from port A and port B, $\mathbf{k}_A$ and $\mathbf{k}_B$, the wavenumber vector of the scattered beam, $\mathbf{k}_S$, and the velocity vector at the measured point, $\mathbf{v}$ [20]. Here, $\mathbf{k}_A$, $\mathbf{k}_B$, and $\mathbf{k}_S$ are all placed in the x-y plane. The incident angle θ_i is defined by the angle between $\mathbf{k}_A$ ($\mathbf{k}_B$) and the x-axis, and the direction of the scattered beam is set to the direction opposite of the x-axis. The beat signal detected by each PD itself provides only the absolute value of its beat frequency. When only the absolute values of the beat frequencies f_A and f_B are known, the direction of velocity cannot be discriminated, but just the magnitudes of the components of the velocity. In order to discriminate the direction of velocity, the signs of beat frequencies f_A and f_B also need to be discriminated. The sign can be discriminated by using the relation between the phases in the signals detected by the two PDs (PD1A and PD2A for the discrimination of the sign of f_A, and PD1B and PD2B for the discrimination of the sign of f_B).

To estimate the characteristics of the proposed LDV for two-dimensional velocity measurement, the relation among velocity vector, beat frequencies, and incident angle θ_i is simulated. Figure 8.7 shows the polar expression of the absolute value of the beat frequency normalized with $|\mathbf{v}/\lambda|$ and the direction of velocity θ_v for various θ_i. The distance from the origin O represents the absolute value of the normalized beat frequency, and the angle from the x-axis represents θ_v, as shown in Figure 8.7e. The sign of the beat frequency for each plot is also indicated in this figure. The proportion between f_A and f_B changes as changing the direction of velocity θ_v with one-to-one correspondence, unless $\theta_i = 0°$. It indicates that the direction of velocity θ_v can be discriminated by the proposed LDV as well as the absolute values of the components of velocity. When the incident angle θ_i increases from $0°$ to $90°$, the plots for f_A and f_B gradually separate each other. It implies that θ_i should be larger for good discrimination of the direction of velocity, although the optimal value of θ_i may also depend on the allowable arrangement of the optical system and the measured point. In the plots for $\theta_i = 0°$ (Figure 8.7a), both beat frequencies f_A and f_B are zero at $\theta_v = 90°$ and $270°$. This indicates that there is no sensitivity to y-directed velocity.

8.3.2   TWO-DIMENSIONAL LDV BY MONITORING BEAMS IN DIFFERENT DIRECTIONS

The method described in Section 8.3.1 uses the discrimination of two orthogonally polarized beams. In this method, polarization cross talk between the polarized beams due mainly to imperfection of polarization-dependent optical components or depolarization on scattering easily results in a measurement error. Hence, a two-dimensional LDV without polarization discrimination is more desirable.

Figure 8.8 illustrates the principle of the proposed LDV for two-dimensional velocity measurement by monitoring beams in different directions [19]. The beam output from a laser is split into four beams, including two beams to be incident on the measured point and two reference beams, with a 1×4 splitter. Two beams are incident on the measured point with different angles and scattered on the object at the measured point. The scattered beams in different directions are monitored with two detection blocks (detection block 1 and detection block 2). In each detection block, the scattered beam is split into three beams with two 1×2 splitters. Two of the three split scattered beams are combined with one of the reference beams split by another 1×2 splitter. Here, the phase of one of the split reference beams is shifted so that the phase difference between two split reference beams becomes $90°$. The power of each combined beam is detected by a PD as a beat signal with three beat frequencies produced by the interference among the scattered beam from port A, the scattered beam from port B, and the reference beam. Two PDs (PD1 and PD2) and the $90°$ phase shift of the reference beam are used to detect the signs of the beat frequencies. The power of the other one of the three split scattered beams is detected by PD3. The beat frequency of the beat signal produced by the interference between the scattered beams from port A and port B can be discriminated by PD3 in each detection block

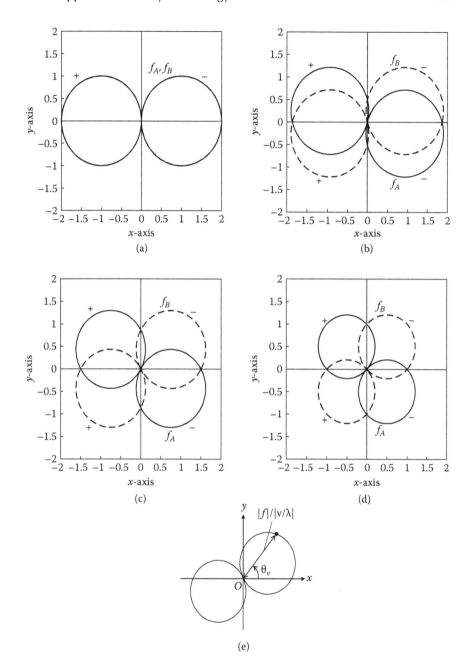

FIGURE 8.7 Polar expression of absolute value of beat frequency normalized with $|\mathbf{v}/\lambda|$ and direction of velocity θ_v for various θ_i. (a) $\theta_i = 0°$, (b) $\theta_i = 30°$, (c) $\theta_i = 60°$, (d) $\theta_i = 90°$, and (e) explanation of plots. The distance from the origin O represents the absolute value of the normalized beat frequency, and the angle from the x-axis represents θ_v. The sign of the beat frequency for each plot is also indicated. (From K. Maru, L. Y. Hu, R. S. Lu, Y. Fujii, and P. P. Yupapin, *Optik*, 122(11), 974–977, 2011.)

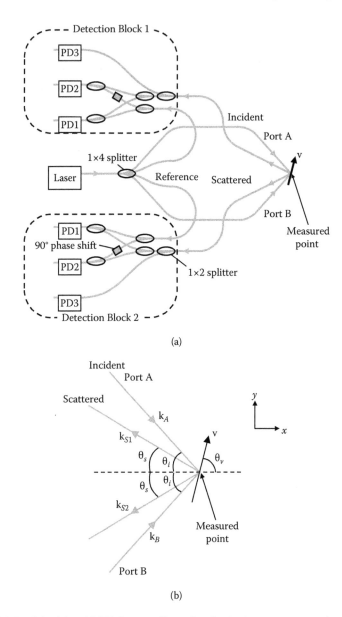

(a)

(b)

FIGURE 8.8 Principle of LDV for two-dimensional velocity measurement by monitoring beams in different directions. (a) Setting of optical system. (b) Diagram of beams at measured point. (From K. Maru and Y. Fujii, *IEEE Sens. J.*, 11(2), 312–318, 2011.)

because of the differential operation of the LDV. In each detection block, the beat frequency produced by the interference between the scattered beam from port A and the reference beam cannot be discriminated from that produced by the interference between the scattered beam from port B and the reference beam. Hence, the direction of velocity cannot be detected only by using one detection block. In order to discriminate the direction of velocity, scattered beams in different directions are monitored with two detection blocks.

The beat frequencies detected with detection block n ($n = 1, 2$), f_{ASn} and f_{BSn}, are derived in a procedure similar to that for the derivation in Section 8.3.1. The principle of the discrimination of velocity direction in the proposed LDV is as follows. In each detection block, the beat frequencies f_{ASn} and f_{BSn} cannot be discriminated from each other, whereas the signs of f_{ASn} and f_{BSn}, as well as their absolute values, can be detected because of the 90° phase shift of the reference beam as described above. Hence, the direction of velocity cannot be discriminated in case only one detection block is used. In the proposed velocimetry, the direction of velocity is discriminated by using two detection blocks in the following principle. The actual velocity vector at the measurement point $\mathbf{v}_r$ can be discriminated by the following procedure even if the two beat frequencies cannot be discriminated with each detection block [19]:

1. Calculate $\mathbf{v}_1$ and $\mathbf{v}_1'$ from the beat frequencies f_1 and f_1' obtained from detection block 1 as the following equations:

$$\mathbf{v}_1 = 2\pi \begin{pmatrix} (\mathbf{k}_{S1}-\mathbf{k}_A)^T \\ (\mathbf{k}_{S1}-\mathbf{k}_B)^T \end{pmatrix}^{-1} \begin{pmatrix} f_1 \\ f_1' \end{pmatrix} \tag{8.2}$$

$$\mathbf{v}_1' = 2\pi \begin{pmatrix} (\mathbf{k}_{S1}-\mathbf{k}_A)^T \\ (\mathbf{k}_{S1}-\mathbf{k}_B)^T \end{pmatrix}^{-1} \begin{pmatrix} f_1' \\ f_1 \end{pmatrix} \tag{8.3}$$

2. Calculate $\mathbf{v}_2$ and $\mathbf{v}_2'$ from the beat frequencies f_2 and f_2' obtained from detection block 2 as the following equations:

$$\mathbf{v}_2 = 2\pi \begin{pmatrix} (\mathbf{k}_{S2}-\mathbf{k}_A)^T \\ (\mathbf{k}_{S2}-\mathbf{k}_B)^T \end{pmatrix}^{-1} \begin{pmatrix} f_2 \\ f_2' \end{pmatrix} \tag{8.4}$$

$$\mathbf{v}_2' = 2\pi \begin{pmatrix} (\mathbf{k}_{S2}-\mathbf{k}_A)^T \\ (\mathbf{k}_{S2}-\mathbf{k}_B)^T \end{pmatrix}^{-1} \begin{pmatrix} f_2' \\ f_2 \end{pmatrix} \tag{8.5}$$

3. When $\mathbf{v}_1$ differs from $\mathbf{v}_1'$, $\mathbf{v}_2$ also differs from $\mathbf{v}_2'$. When $\mathbf{v}_1$ corresponds to $\mathbf{v}_2$ or $\mathbf{v}_2'$, $\mathbf{v}_1$ is the same as $\mathbf{v}_r$. When $\mathbf{v}_1'$ corresponds to $\mathbf{v}_2$ or $\mathbf{v}_2'$, $\mathbf{v}_1'$ is the same as $\mathbf{v}_r$.
4. When $\mathbf{v}_1$ corresponds to $\mathbf{v}_1'$, $\mathbf{v}_2$ also corresponds to $\mathbf{v}_2'$, and all the four vectors corresponds to $\mathbf{v}_r$.

Here, $\mathbf{k}_A$ and $\mathbf{k}_B$ are the wavenumber vectors of the incident beam from port A and port B to each block, respectively, and $\mathbf{k}_{Sn}$ is the wavenumber vector of the scattered beam detected with detection block n. The incident angle θ_i is defined by the angle between $\mathbf{k}_A$ ($\mathbf{k}_B$) and the x-axis, and the angle θ_s is defined by the angle between $\mathbf{k}_{S1}$ ($\mathbf{k}_{S2}$) and the direction opposite to the x-axis, as shown in Figure 8.8b.

Let $\mathbf{v}_{i1}$ be defined as either of $\mathbf{v}_1$ or $\mathbf{v}_1'$, which does not correspond to $\mathbf{v}_r$, and $\mathbf{v}_{i2}$ be defined as either of $\mathbf{v}_2$ or $\mathbf{v}_2'$, which does not correspond to $\mathbf{v}_r$. Provided that we consider the transformation from $\mathbf{v}_r$ to $\mathbf{v}_{i1}$, there are two eigenvectors $\mathbf{x}_{S1,1}$ and $\mathbf{x}_{S1,-1}$ with the eigenvalues of 1 and -1, respectively. Similarly, there are two eigenvectors $\mathbf{x}_{S2,1}$ and $\mathbf{x}_{S2,-1}$ with the eigenvalues of 1 and -1 on the transformation from $\mathbf{v}_r$ to $\mathbf{v}_{i2}$. This implies that when the actual velocity vector $\mathbf{v}_r$ is directed to $\mathbf{x}_{S1,1}$ (or $\mathbf{x}_{S2,1}$), both $\mathbf{v}_{i1}$ and $\mathbf{v}_{i2}$ become the same as $\mathbf{v}_r$. On the other hand, when $\mathbf{v}_r$ is directed to the different direction from $\mathbf{x}_{S1,1}$ (or $\mathbf{x}_{S2,1}$), $\mathbf{v}_r$ has the components of $\mathbf{x}_{S1,-1}$ and $\mathbf{x}_{S2,-1}$. Here, the eigenvectors $\mathbf{x}_{S1,-1}$ and $\mathbf{x}_{S2,-1}$ are different from each other because the wavenumber vectors $\mathbf{k}_{S1}$ and $\mathbf{k}_{S2}$ are different from each other, whereas their eigenvalues are the same. This means that the resultant vectors of these components are different from each other even after the transformations.

Figure 8.9 shows examples of the directional relation among $\mathbf{v}_r$, $\mathbf{v}_{i1}$, and $\mathbf{v}_{i2}$ for various directions of the velocity, θ_{vr}, defined as the angle between the direction of $\mathbf{v}_r$ and the x-axis. The magnitude of $\mathbf{v}_r$ is normalized to unity, and it is assumed that $\theta_i = 60°$ and $\theta_s = 50°$. In this case, the eigenvectors $\mathbf{x}_{S1,1}$ and $\mathbf{x}_{S2,1}$ are directed to the parallel direction to the x-axis. Consequently, both $\mathbf{v}_{i1}$ and $\mathbf{v}_{i2}$ correspond to $\mathbf{v}_r$ when θ_{vr} is $0°$ or $180°$. When the θ_{vr} is $45°$, $90°$, or $135°$, $\mathbf{v}_{i1}$ and $\mathbf{v}_{i2}$ are different from each other, and both $\mathbf{v}_{i1}$ and $\mathbf{v}_{i2}$ are also different from $\mathbf{v}_r$. It is because the eigenvectors $\mathbf{x}_{S1,-1}$ and $\mathbf{x}_{S2,-1}$ are directed to different directions ($56.2°$ and $123.8°$ from the x-axis, respectively), and both the $\mathbf{x}_{S1,-1}$ component and the $\mathbf{x}_{S2,-1}$ component of $\mathbf{v}_r$ are oppositely directed after the transformation.

Figure 8.10 shows the magnitudes and directions of the velocities $\mathbf{v}_r$, $\mathbf{v}_{i1}$, and $\mathbf{v}_{i2}$ as a function of the direction of the velocity θ_{vr}. Like in Figure 8.9, the magnitude of $\mathbf{v}_r$ is normalized to unity, and it is assumed that $\theta_i = 60°$ and $\theta_s = 50°$. Both $\mathbf{v}_{i1}$ and $\mathbf{v}_{i2}$ correspond to $\mathbf{v}_r$ only at $\theta_{vr} = 0°$ or $180°$. When the θ_{vr} is different from $0°$ or $180°$, $\mathbf{v}_{i1}$ and $\mathbf{v}_{i2}$ are different from each other, and both $\mathbf{v}_{i1}$ and $\mathbf{v}_{i2}$ are also different from $\mathbf{v}_r$. In the proposed structure, $\mathbf{v}_{i1}$ is obtained from detection block 1, $\mathbf{v}_{i2}$ is obtained from detection block 2, and $\mathbf{v}_r$ is obtained from both of the detection blocks. Consequently, for all the directions of $\mathbf{v}_r$, the velocity vector $\mathbf{v}_r$ can be discriminated as one of two velocity vectors obtained from detection block 1, which corresponds to one of two velocity vectors obtained from detection block 2.

8.4 SUMMARY

Several types of integrated LDVs using the PLC technology, especially wavelength-insensitive LDVs and a multipoint LDV using AWGs, are described. The PLC technology has a potential for drastically reducing the sizes of LDVs. By optimizing the design parameters, a wavelength-insensitive operation can be obtained by using AWGs without assembly of a bulk diffractive grating and other optical elements. The use of AWGs also has the potential to simultaneously measure velocities at different points in compact optical systems.

In addition, techniques for two-dimensional velocity measurement using a simple optical configuration without multiple colors or any optical modulator are described.

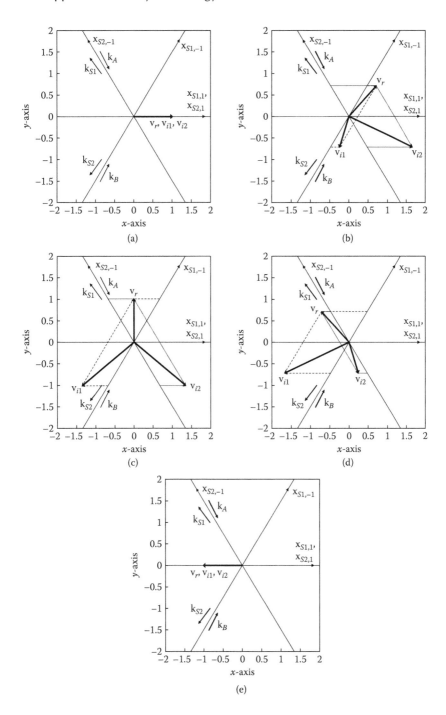

FIGURE 8.9 Directional relation among $\mathbf{v}_r$, $\mathbf{v}_{i1}$, and $\mathbf{v}_{i2}$ at $\theta_i = 60°$ and $\theta_s = 50°$ for (a) $\theta_{vr} = 0°$, (b) $\theta_{vr} = 45°$, (c) $\theta_{vr} = 90°$, (d) $\theta_{vr} = 135°$, and (e) $\theta_{vr} = 180°$. (From K. Maru and Y. Fujii, *IEEE Sens. J.*, 11(2), 312–318, 2011.)

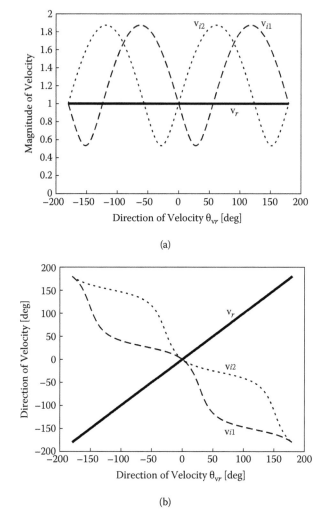

FIGURE 8.10 Magnitudes and directions of $\mathbf{v}_r$, $\mathbf{v}_{i1}$, and $\mathbf{v}_{i2}$ as a function of direction of velocity θ_{vr}: (a) magnitudes and (b) directions. $\theta_i = 60°$ and $\theta_s = 50°$. The magnitude of $\mathbf{v}_r$ is normalized to unity. (From K. Maru and Y. Fujii, *IEEE Sens. J.*, 11(2), 312–318, 2011.)

In the proposed methods, the combination of polarized beams and the 90° phase shift or the monitoring of the scattered beams in different directions with two detection blocks is used to discriminate the direction of velocity. In each method, the simulation result indicates that the two-dimensional velocity vector can be discriminated. These methods would be applicable to various velocity measurement fields.

REFERENCES

1. Y. Yeh and H. Z. Cummins, Localized fluid flow measurements with an He-Ne laser spectrometer, *Appl. Phys. Lett.*, 4(10), 176–178, 1964.

2. M. Kawachi, Silica waveguides on silicon and their application to integrated-optic components, *Optical Quantum Electron.*, 22, 391–416, 1990.
3. H.-E. Albrecht, M. Borys, N. Damaschke, and C. Tropea, *Laser Doppler and phase Doppler measurement techniques*, Springer-Verlag, Berlin, 2003, chap. 7.
4. K. A. Blake, Simple two-dimensional laser velocimeter optics, *J. Phys. E*, 5, 623–624, 1972.
5. N. Nakatani, M. Tokita, and T. Yamada, LDV using polarization-preserving optical fibers for simultaneous measurement of two velocity components, *Appl. Opt.*, 23(11), 1686–1687, 1984.
6. N. Nakatani, M. Tokita, T. Izumi, and T. Yamada, Laser Doppler velocimetry using polarization-preserving optical fibers for simultaneous measurement of multidimensional velocity components, *Rev. Sci. Instrum.*, 56, 2025–2029, 1985.
7. G. T. Grant and K. T. Orloff, Two-color dual beam backscatter laser Doppler velocimeter, *Appl. Opt.*, 12(12), 2913–2916, 1973.
8. H.-E. Albrecht, M. Borys, N. Damaschke, and C. Tropea, *Laser Doppler and phase Doppler measurement techniques*, Springer-Verlag, Berlin, 2003, Section 7.4.2.
9. R. J. Adrian, A bipolar, two component laser-Doppler velocimeter, *J. Phys. E*, 8, 723–726, 1975.
10. X. Shen, J. Zhang, Z. Wang, and H. Yu, Two component LDV system with dual-differential acousto-optical frequency shift and its applications, *Acta Mech Sin*, 2, 81–92, 1986.
11. S. Kato and K. Hasegawa, Novel techniques of laser Doppler velocimetry using optical integrated circuit, *R&D Rev. Toyota Central R&D Labs.*, 34, 35–42, 1999 (in Japanese).
12. K. Hasegawa, S. Kato, and H. Itoh, Two-dimensional fiber laser Doppler velocimeter by integrated optical frequency shifter, *Proc. SPIE*, 3740, 294–297, 1999.
13. L. Kervevan, H. Gilles, S. Girard, and M. Laroche, Two-dimensional velocity measurements with self-mixing technique in diode-pumped Yb:Er glass laser, *IEEE Photon. Technol. Lett.*, 16(7), 1709–1711, 2004.
14. K. Maru and Y. Fujii, Wavelength-insensitive laser Doppler velocimeter using beam position shift induced by Mach-Zehnder interferometers, *Opt. Express*, 17(20), 17441–17449, 2009.
15. K. Maru and Y. Fujii, Integrated wavelength-insensitive differential laser Doppler velocimeter using planar lightwave circuit, *J. Lightwave Technol.*, 27(22), 5078–5083, 2009.
16. K. Maru and Y. Fujii, Differential laser Doppler velocimeter with enhanced range for small wavelength sensitivity by using cascaded Mach-Zehnder interferometers, *J. Lightwave Technol.*, 28(11), 1631–1637, 2010.
17. K. Maru and Y. Fujii, Laser Doppler velocimetry with small wavelength sensitivity using planar lightwave circuit, *Appl. Mechanics Mater.*, 103, 76–81, 2011.
18. K. Maru, K. Kobayashi, and Y. Fujii, Multi-point differential laser Doppler velocimeter using arrayed waveguide gratings with small wavelength sensitivity, *Opt. Express*, 18(1), 301–308, 2010.
19. K. Maru and Y. Fujii, Laser Doppler velocimetry for two-dimensional directional discrimination by monitoring scattered beams in different directions, *IEEE Sens. J.*, 11(2), 312–318, 2011.
20. K. Maru, L. Y. Hu, R. S. Lu, Y. Fujii, and P. P. Yupapin, Two-dimensional laser Doppler velocimeter using polarized beams and 90° phase shift for discrimination of velocity direction, *Optik*, 122(11), 974–977, 2011.
21. M. K. Smit, New focusing and dispersive planar component based on an optical phased array, *Electron. Lett.*, 24(7), 385–386, 1988.
22. C. Dragone, Efficient N×N star coupler based on Fourier optics, *Electron. Lett.*, 24(15), 942–944, 1988.

23. H. Takahashi, S. Suzuki, K. Kato, and I. Nishi, Arrayed-waveguide grating for wavelength division multi/demultiplexer with nanometer resolution, *Electron. Lett.*, 26(2), 87–88, 1990.
24. H. Takahashi, I. Nishi, and Y. Hibino, 10 GHz spacing optical frequency division multiplexer based on arrayed-waveguide grating, *Elecron. Lett.*, 28(4), 380–382, 1992.
25. A. R. Vellekoop and M. K. Smit, Four-channel integrated-optic wavelength demultiplexer with weak polarization dependence, *J. Lightwave Technol.*, 9(3), 310–314, 1991.
26. C. Dragone, An N × N optical multiplexer using a planar arrangement of two star couplers, *IEEE Photon. Technol. Lett.*, 3, 812–815, 1991.
27. C. Dragone, C. A. Edwards, and R. C. Kistler, Integrated optics N × N multiplexer on silicon, *IEEE Photon. Technol. Lett.*, 3, 896–899, 1991.
28. Y. Tachikawa, Y. Inoue, M. Kawachi, H. Takahashi, and K. Inoue, Arrayed-waveguide grating add-drop multiplexer with loop-back optical paths, *Electron. Lett.*, 29(24), 2133–2134, 1993.
29. O. Ishida, H. Takahashi, S. Suzuki, and Y. Inoue, Multichannel frequency-selective switch employing an arrayed-waveguide grating multiplexer with fold-back optical paths, *IEEE Photon. Technol. Lett.*, 6(10), 1219–1221, 1994.
30. M. Kawachi, M. Yasu, and M. Kobayashi, Flame hydrolysis deposition of SiO_2-TiO_2 glass planar optical waveguides on silicon, *Jpn. J. Appl. Phys.*, 22(12), 1932, 1983.
31. S. Kashimura, M. Takeuchi, K. Maru, and H. Okano, Loss reduction of GeO_2-doped silica waveguide with high refractive index difference by high-temperature annealing, *Jpn. J. Appl. Phys.*, 39(6A), L521–L523, 2000.
32. C. H. Henry, G. E. Blonder, and R. F. Kazarinov, Glass waveguides on silicon for hybrid optical packaging, *J. Lightwave Technol.*, 7(10), 1530–1539, 1989.
33. C. R. Doerr and K. Okamoto, Advances in silica planar lightwave circuits, *J. Lightwave Technol.*, 24(12), 4763–4789, 2006.
34. J. Schmidt, R. Volkel, W. Stork, J. T. Sheridan, J. Schwider, and N. Steibl, Diffractive beam splitter for laser Doppler velocimetry, *Opt. Lett.*, 17(17), 1240–1242, 1992.
35. R. Sawada, K. Hane, and E. Higurashi, Optical micro electro mechanical systems, *Ohmsha* (Tokyo), 2002, Section 5.2 (in Japanese).
36. H.-E. Albrecht, M. Borys, N. Damaschke, and C. Tropea, *Laser Doppler and phase Doppler measurement techniques*, Springer-Verlag, Berlin, 2003, Section 7.2.2.
37. T. Hachiga, N. Furuichi, J. Mimatsu, K. Hishida, and M. Kumada, Development of a multi-point LDV by using semiconductor laser with FFT-based multi-channel signal processing, *Exp. Fluids*, 24, 70–76, 1998.
38. M. Haruna, K. Kasazumi, and H. Nishihara, Integrated-optic differential laser Doppler velocimeter with a micro Fresnel lens array, in *Proceedings of the Conference on Integrated and Guided-Wave Optics (IGWO '89)*, New York, N.Y., p. MBB6.
39. T. Ito, R. Sawada, and E Higurashi, Integrated microlaser Doppler velocimeter, *J. Lightwave Technol.*, 17(1), 30–34, 1999.
40. C. R. Doerr, M. Cappuzzo, E. Laskowski, A. Paunescu, L. Gomez, L. W. Stulz, and J. Gates, Dynamic wavelength equalizer in silica using the single-filtered-arm interferometer, *IEEE Photon. Technol. Lett.*, 11(5), 581–583, 1999.
41. K. Plamann, H. Zellmer, J. Czarske, and A. Tünnermann, Directional discrimination in laser Doppler anemometry (LDA) without frequency shifting using twinned optical fibres in the receiving optics, *Meas. Sci. Technol.*, 9, 1840–1846, 1998.

9 Differential Photosensors for Optical Wireless Communication and Location Technologies

Xian Jin, Ahmed Arafa, Blago A. Hristovski, Richard Klukas, and Jonathan F. Holzman

CONTENTS

9.1 INTRODUCTION AND BACKGROUND

Sensing is the basis for probing external environments, and improvements in sensing have followed the development of various signal emission and detection technologies. Sensor technologies gather information from distributed targets by way of two general formats: passive sensing and active sensing.

The first format, passive sensing,[1-3] is unidirectional in nature, as information is simply gathered from the external environment. Natural or artificial targets emit visible, infrared (IR), or radio frequency (RF) signals, and these signals are detected by the passive sensor network. Charge-coupled devices (CCD)[4] and complementary metal oxide semiconductor (CMOS) sensors[5] are classic examples of such a passive system for visible/IR spectra. The global positioning system (GPS) is an example of passive reception in the RF regime.[6,7] Ultimately, the selected technology must be appropriate for the scale of the desired detection environment, with long-wavelength RF systems being well suited to global scales and short-wavelength optical systems being better suited to applications requiring high sensitivities and small spatial resolutions.

The second format, active sensing,[8–10] differs from its passive counterpart through its increased level of control and bidirectional implementation. Active sensing can effectively probe its external environment by emitting signals and sampling their reflected/backscattered light level, as the source characteristics can be controlled (through modulation, polarization, wavelength, etc.) to improve the signal sensitivity. This principle has been successfully applied in long-wavelength remote-sensing applications such as millimeter-wave radar,[11] short-range radar,[12] and even mid-IR terahertz (THz) imaging[13] and radar ranging.[14] Technologies based upon active sensing are particularly effective for use in multidirectional sensor links.

An intriguing application for multidirectional active sensing exists in modern smart dust technologies.[15] These technologies establish bidirectional optical links between multiple sensor nodes. Two-way communication is established between each sensor node through the realization of three distinct tasks: photodetection of the incident light, retroreflection of the incident light, and modulation of the retroreflected light. Photodetection is typically accomplished with a photodiode (PD), while retroreflection is provided by a separate corner-cube retroreflector. Modulation of the returning signal, for communication in the bidirectional link, is typically accomplished by mechanical beam deflectors on the incorporated corner-cube mirrors,[16] liquid crystal (LC) shutters,[17] or multiple quantum well modulators.[18]

Given these emerging multidirectional technologies, we introduce here a differential photosensor, which merges the photodetection and retroreflection capabilities of active sensing. This photosensor is composed of three mutually orthogonal PDs in a corner-cube architecture. Unlike standard PD[19] and photoconductive[20] switching technologies, which record a single magnitude for the incident optical power, such a structure allows for simultaneous retroreflection of incident signals back to their source (through the retroreflecting nature of the corner-cube geometry) and photodetection of the incident power level (through the parallel sum of the individual photocurrents). The three-channel PD structure also provides a means to quantify the optical alignment. Indeed, differential combinations of PD photocurrents can be used to triangulate the direction of the optical source and align the structure along the optimal orientation. This can support optical sensing for optical wireless communication (OWC) technology as well as optical wireless location (OWL) technology. The latter, positioning technology, can be especially effective for the indoor environment, where GPS signals have severe limitations due to attenuation and multipath reflections, and inertial navigation system (INS) signals exhibit drift due to the accumulated errors from accelerometer and gyroscope data.

In this chapter, the proposed differential photosensor is introduced by way of its design, theory, and applications to OWC and OWL technologies.

9.2   SYSTEM DESIGN AND OPERATION

The proposed photosensor is assembled as the corner-cube structure shown in Figure 9.1a. Silicon PDs with the uniform and smooth surfaces are selected to provide enhanced reflectivity with minimal surface scattering/diffraction. Each of them has a 9.7×9.7 mm^2 electrically isolated active area with negligible inactive edges around the periphery. The PDs are bonded into a mutually orthogonal corner-cube form during an alignment/calibration process, with PD_1 lying in the yz-plane, PD_2 lying in the xz-plane, and PD_3 lying in the xy-plane. To ensure the required retroreflection accuracy, the retroreflection is monitored over a 5 m length while the unit is bonded. A high-speed Pi-cell LC optical modulator is bonded to the entrance interface of the photosensor to modulate the incident/retroreflected optical beams.

The orthogonal nature of the photosensor provides retroreflection, which allows incident light to be reflected directly back to its source. This is accomplished through three internal reflections off of the three constituent PDs. The use of PDs here also offers an opportunity to probe the power levels of incident light by sampling the

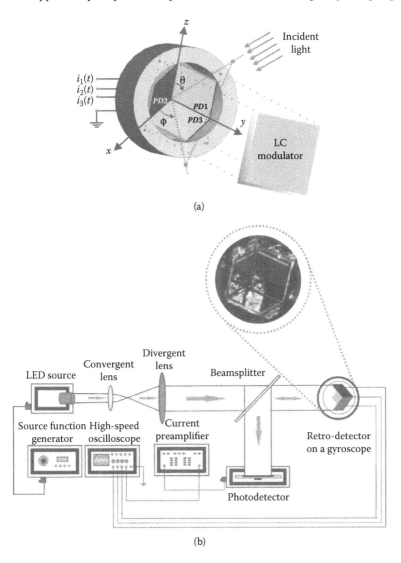

FIGURE 9.1 (See color insert.) (a) Differential photosensor. (b) OWC system with a LED as the light source. The photosensor is shown in the inset. *(continued)*

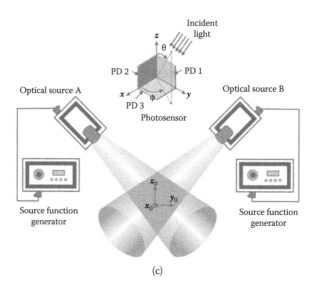

(c)

FIGURE 9.1 (CONTINUED)   (c) OWL system with two optical sources.

beam power. The three resulting photocurrents are recorded as $i_1(t)$, $i_2(t)$, and $i_3(t)$, as shown in Figure 9.1a, for PD_1, PD_2, and PD_3, respectively. The sum of these photocurrents provides a measure of the total incoming optical signal power level. The constituent photocurrents of the photosensor provide a means to triangulate the orientations for incident optical beams, by way of the azimuthal φ and polar θ angles shown in Figure 9.1a.

9.3   THEORY

The geometrical nature of the differential photosensor allows it to act as a retroreflector, by reflecting a fraction of the incident light back to its source, and a photodetector, by sampling the incident light with the PDs. The theory for the resulting retroreflection and photodetection characteristics is presented within the following subsections.

9.3.1   Retroreflection Characteristics

Active sensing targets must effectively redirect incident light back to their respective sources. A suitable structure for accomplishing this retroreflection is the standard corner-cube retroreflector. Such an architecture consists of three reflective surfaces being mutually perpendicular, as shown in Figure 9.1a. When incident light enters this corner along the unit-normal vector,

$$\hat{r} = -n_1\hat{x} - n_2\hat{y} - n_3\hat{z} \qquad (9.1)$$

it undergoes a series of reflections, reversing the n_1, n_2, and n_3 incident ray directional cosines in the x-, y-, and z-directions, respectively. The uniform incident intensity,

$$I_0 \hat{r} = -I_0 n_1 \hat{x} - I_0 n_2 \hat{y} - I_0 n_3 \hat{z} \tag{9.2}$$

for a uniform intensity magnitude of I_0, leads to intensities of

$$\bar{I}_1 = +RI_0 n_1 \hat{x} - RI_0 n_2 \hat{y} - RI_0 n_3 \hat{z}$$

$$\bar{I}_2 = -RI_0 n_1 \hat{x} + RI_0 n_2 \hat{y} - RI_0 n_3 \hat{z} \tag{9.3}$$

$$\bar{I}_3 = -RI_0 n_1 \hat{x} - RI_0 n_2 \hat{y} + RI_0 n_3 \hat{z}$$

after the light undergoes reflections from PD_1, PD_2, and PD_3, respectively. The angular-independent surface reflectivity here is denoted by R. After three successful internal reflections, light exits the corner-cube antiparallel to the incident ray with an intensity defined by

$$-I_0 R^3 \hat{r} = +I_0 R^3 n_1 \hat{x} + I_0 R^3 n_2 \hat{y} + I_0 R^3 n_3 \hat{z} \tag{9.4}$$

The degree to which this structure acts as a retroreflector can be quantified by the reflection solid angle subtended by the azimuthal φ and polar θ angles. The definitions for these angles are shown in Figure 9.1 along with the directional cosine components:

$$n_1 = \cos(\varphi) \cdot \sin(\theta)$$

$$n_2 = \sin(\varphi) \cdot \sin(\theta) \tag{9.5}$$

$$n_3 = \cos(\theta)$$

The retroreflective response of this corner-cube structure has been successfully applied to short-[21] and long[22]-range bidirectional applications, and the angular analyses for the retroreflection directionality are shown elsewhere.[23]

9.3.2 Photodetection Characteristics

The desired photodetection process for the proposed photosensor is based on the use of three orthogonal silicon PDs. The incident light with a uniform intensity illuminates the element shown in Figure 9.1a, and this (typically) leads to three PD photocurrents with differing amplitudes. The disparity between these amplitudes is due to imbalances between the incident light ray directional cosine components. Each PD will have a differing cross-sectional illumination area as viewed from the source: incident light with a large n_1 component along the x-axis will preferentially illuminate PD_1, incident light with a large n_2 component along the y-axis will preferentially illuminate PD_2, and incident light with a large n_3 component along the z-axis will preferentially illuminate PD_3. Quantifying this relationship becomes further complicated by the fact that the recorded PD photocurrents will also have contributions from light rays after primary and secondary internal reflections. The complete theoretical model for the PD photocurrents, including these internal reflections, is developed in this section. Power levels incident upon PD_1, PD_2, and PD_3 are investigated as three individual responses with results defined in terms of six illumination condition cases:

1. $n_1 < n_2 < n_3$
2. $n_1 < n_3 < n_2$
3. $n_2 < n_1 < n_3$
4. $n_2 < n_3 < n_1$
5. $n_3 < n_1 < n_2$
6. $n_3 < n_2 < n_1$

Each of these cases is itself characterized by six individual subcases, where the photocurrent levels depend upon the ordering of the primary and secondary reflections. Fortunately, arguments based on rotational and mirror symmetries can be employed, and the required analysis is demonstrated here specifically for the $n_1 < n_2 < n_3$ case.

The first analysis case corresponds to incident illumination with the x, y, and z directional cosine components in increasing order of magnitudes. This $n_1 < n_2 < n_3$ situation is analyzed for the first subcase, in which the incident light strikes PD_1, reflects onto PD_2, and reflects onto PD_3 (and subsequently exits the photosensor to return to its source). In this case, the light ray unit-normal vectors that are incident on PD_1, PD_2, and PD_3 are

$$\hat{r}_1 = -n_1\hat{x} - n_2\hat{y} - n_3\hat{z}$$

$$\hat{r}_{12} = +n_1\hat{x} - n_2\hat{y} - n_3\hat{z} \tag{9.6}$$

$$\hat{r}_{123} = +n_1\hat{x} + n_2\hat{y} - n_3\hat{z}$$

respectively, where the subscripts indicate the successive illumination and reflection sequence. Figure 9.2a shows the resulting illumination areas for this subcase. Here, PD_1 is fully illuminated and shaded, and its four corners in the yz-plane are projected along $\hat{r}_{12} = +n_1\hat{x} - n_2\hat{y} - n_3\hat{z}$ onto the xz-plane. The resulting shaded PD_2 illumination area is then defined as the overlap between this projected area and the PD_2 surface. Similarly, the PD_2 illumination area is projected along $\hat{r}_{123} = +n_1\hat{x} + n_2\hat{y} - n_3\hat{z}$ onto the xy-plane, and the shaded PD_3 illumination is defined as the overlap between this projected area and the PD_3 surface. The resulting illumination area normal vectors can now be defined for this subcase as

$$\vec{A}_1 = a^2\hat{x}$$

$$\vec{A}_{12} = a^2 \frac{n_1}{2n_3}\hat{y} \tag{9.7}$$

$$\vec{A}_{123} = a^2 \frac{n_1 n_2}{2n_3^2}\hat{z}$$

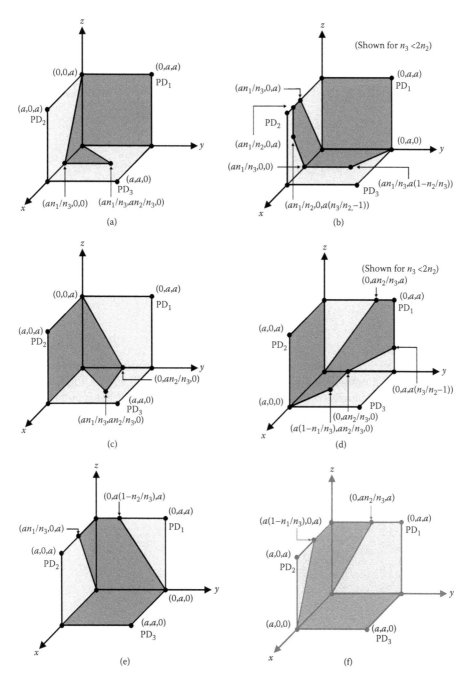

FIGURE 9.2   Schematics are shown for the internal reflection processes occurring in the photosensor for directional cosine conditions $n_1 < n_2 < n_3$. Nonilluminated areas are shown with light shading. Illuminated areas are shown with dark shading. The order of internal reflections proceeds as (a) PD_1 onto PD_2 onto PD_3, (b) PD_1 onto PD_3 onto PD_2, (c) PD_2 onto PD_1 onto PD_3, (d) PD_2 onto PD_3 onto PD_1, (e) PD_3 onto PD_1 onto PD_2, and (f) PD_3 onto PD_2 onto PD_1.

The corresponding incident power levels associated with PD_1, PD_2, and PD_3 are then found by taking the component of these illumination area normal vectors in (9.7) along the respective reflected light ray unit-normal vectors in (9.6). The result is

$$P_1 = -I_0 \hat{r}_1 \cdot \vec{A}_1 = I_0 a^2 n_1$$

$$P_{12} = -RI_0 \hat{r}_{12} \cdot \vec{A}_{12} = RI_0 a^2 \frac{n_1 n_2}{2n_3}$$

$$P_{123} = -R^2 I_0 \hat{r}_{123} \cdot \vec{A}_{123} = R^2 I_0 a^2 \frac{n_1 n_2}{2n_3}$$

$$(9.8)$$

The second subcase for $n_1 < n_2 < n_3$ corresponds to the situation with the incident light illuminating PD_1, PD_3, and PD_2, in order. In this case, the respective light ray unit-normal vectors incident on the surfaces of PD_1, PD_3, and PD_2 are

$$\hat{r}_1 = -n_1 \hat{x} - n_2 \hat{y} - n_3 \hat{z}$$

$$\hat{r}_{13} = +n_1 \hat{x} - n_2 \hat{y} - n_3 \hat{z}$$

$$\hat{r}_{132} = +n_1 \hat{x} - n_2 \hat{y} + n_3 \hat{z}$$

$$(9.9)$$

Again, the successive illumination areas are referred onto their adjoining neighbors, and the illumination areas are displayed in Figure 9.2b (shown for $n_3 < 2n_2$). Using the shaded areas on this figure, the illumination area normal vectors can be defined as

$$\vec{A}_1 = a^2 \hat{x}$$

$$\vec{A}_{13} = a^2 \left(\frac{n_1}{n_3} - \frac{n_1 n_2}{2n_3^2} \right) \hat{z}$$

$$\vec{A}_{132} = a^2 \begin{cases} \left(2\dfrac{n_1}{n_2} - \dfrac{n_1}{n_3} - \dfrac{n_1 n_3}{2n_2^2} \right) \hat{y}, n_3 < 2n_2 \\[2ex] \dfrac{n_1}{n_3} \hat{y}, n_3 > 2n_2 \end{cases}$$

$$(9.10)$$

The corresponding incident power levels associated with PD_1, PD_3, and PD_2 are then found by taking the component of these illumination area normal vectors, from (9.10), along the respective reflected light ray unit-normal vectors, from (9.9). The result is

$$P_1 = -I_0\hat{r}_1 \cdot \vec{A}_1 = I_0 a^2 n_1$$

$$P_{13} = -RI_0\hat{r}_{13} \cdot \vec{A}_{13} = RI_0 a^2 \left(n_1 - \frac{n_1 n_2}{2n_3} \right)$$

$$(9.11)$$

$$P_{132} = -R^2 I_0 \hat{r}_{132} \cdot \vec{A}_{132} = R^2 I_0 a^2 \begin{cases} \left(2n_1 - \dfrac{n_1 n_2}{n_3} - \dfrac{n_1 n_3}{2n_2} \right), n_3 < 2n_2 \\[2ex] \dfrac{n_1 n_2}{n_3}, n_3 > 2n_2 \end{cases}$$

The third subcase for $n_1 < n_2 < n_3$ is shown in Figure 9.2c, and it is character-ized by illumination of PD_2, followed by subsequent reflections off PD_1 and PD_3. The light ray unit-normal vectors incident on the surfaces of PD_2, PD_1, and PD_3, respectively, are

$$\hat{r}_2 = -n_1\hat{x} - n_2\hat{y} - n_3\hat{z}$$

$$\hat{r}_{21} = -n_1\hat{x} + n_2\hat{y} - n_3\hat{z}$$

$$(9.12)$$

$$\hat{r}_{213} = +n_1\hat{x} + n_2\hat{y} - n_3\hat{z}$$

Using these light ray components and the resulting illumination area projections, the respective illumination area normal vectors can be written for PD_2, PD_1, and PD_3 as

$$\vec{A}_2 = a^2\hat{y}$$

$$\vec{A}_{21} = a^2 \frac{n_2}{2n_3}\hat{x}$$

$$(9.13)$$

$$\vec{A}_{213} = a^2 \frac{n_1 n_2}{2n_3^2}\hat{z}$$

The resulting incident power levels associated with PD_2, PD_1, and PD_3 are then found by taking the component of the illumination area normal vectors, from (9.13), along the respective reflected light ray unit-normal vectors, from (9.12), such that

$$P_2 = -I_0\hat{r}_2 \cdot \vec{A}_2 = I_0 a^2 n_2$$

$$P_{21} = -RI_0\hat{r}_{21} \cdot \vec{A}_{21} = RI_0 a^2 \frac{n_1 n_2}{2n_3}$$

$$(9.14)$$

$$P_{213} = -R^2 I_0\hat{r}_{213} \cdot \vec{A}_{213} = R^2 I_0 a^2 \frac{n_1 n_2}{2n_3}$$

The fourth subcase for $n_1 < n_2 < n_3$ is shown in Figure 9.2d, and it corresponds to the case for which PD_2, PD_3, and PD_1 are successively illuminated. The respective light ray unit-normal vectors incident on PD_2, PD_3, and PD_1 can be expressed as

$$\hat{r}_2 = -n_1\hat{x} - n_2\hat{y} - n_3\hat{z}$$

$$\hat{r}_{23} = -n_1\hat{x} + n_2\hat{y} - n_3\hat{z} \tag{9.15}$$

$$\hat{r}_{231} = -n_1\hat{x} + n_2\hat{y} + n_3\hat{z}$$

and the PD_2, PD_3, and PD_1 illumination area normal vectors are

$$\vec{A}_2 = a^2\hat{y}$$

$$\vec{A}_{23} = a^2\left(\frac{n_2}{n_3} - \frac{n_1 n_2}{2n_3^2}\right)\hat{z}$$

$$\vec{A}_{231} = a^2 \begin{cases} \left(2 - \dfrac{n_2}{n_3} - \dfrac{n_3}{2n_2}\right)\hat{x}, n_3 < 2n_2 \\[2mm] \dfrac{n_2}{n_3}\hat{x}, n_3 > 2n_2 \end{cases} \tag{9.16}$$

The PD_2, PD_3, and PD_1 incident powers are then found by taking the illumination area normal vector components, from (9.16), along the reflected light ray unit-normal vectors, from (9.15), giving

$$P_2 = -I_0\hat{r}_2 \cdot \vec{A}_2 = I_0 a^2 n_2$$

$$P_{23} = -RI_0\hat{r}_{23} \cdot \vec{A}_{23} = RI_0 a^2\left(n_2 - \frac{n_1 n_2}{2n_3}\right)$$

$$P_{231} = -R^2 I_0\hat{r}_{231} \cdot \vec{A}_{231} = R^2 I_0 a^2 \begin{cases} \left(2n_1 - \dfrac{n_1 n_2}{n_3} - \dfrac{n_1 n_3}{2n_2}\right), n_3 < 2n_2 \\[2mm] \dfrac{n_1 n_2}{n_3}, n_3 > 2n_2 \end{cases} \tag{9.17}$$

The fifth subcase for $n_1 < n_2 < n_3$ has illumination and reflection conditions shown in Figure 9.2e. Here, light rays are reflected in order off of PD_3, PD_1, and PD_2. The light ray unit-normal vectors incident on the surfaces of PD_3, PD_1, and PD_2 become

$$\hat{r}_3 = -n_1\hat{x} - n_2\hat{y} - n_3\hat{z}$$

$$\hat{r}_{31} = -n_1\hat{x} - n_2\hat{y} + n_3\hat{z} \tag{9.18}$$

$$\hat{r}_{312} = +n_1\hat{x} - n_2\hat{y} + n_3\hat{z}$$

and the resulting illumination area normal vectors for illumination of PD_3, PD_1, and PD_2 are

$$\vec{A}_3 = a^2 \hat{z}$$

$$\vec{A}_{31} = a^2 \left(1 - \frac{n_2}{2n_3}\right)\hat{x} \tag{9.19}$$

$$\vec{A}_{312} = a^2 \frac{n_1}{2n_3}\hat{y}$$

The resulting incident power levels recorded by PD_3, PD_1, and PD_2 for this subcase are then found, from (9.18) and (9.19), to be

$$P_3 = -I_0 \hat{r}_3 \cdot \vec{A}_3 = I_0 a^2 n_3$$

$$P_{31} = -RI_0 \hat{r}_{31} \cdot \vec{A}_{31} = RI_0 a^2 \left(n_1 - \frac{n_1 n_2}{2n_3}\right) \tag{9.20}$$

$$P_{312} = -R^2 I_0 \hat{r}_{312} \cdot \vec{A}_{312} = R^2 I_0 a^2 \frac{n_1 n_2}{2n_3}$$

The sixth and final subcase for $n_1 < n_2 < n_3$ is displayed in Figure 9.2f. The surface of PD_3 is illuminated first, followed by successive reflections off of PD_2 and PD_1. The light ray unit-normal vectors for this succession are

$$\hat{r}_3 = -n_1 \hat{x} - n_2 \hat{y} - n_3 \hat{z}$$

$$\hat{r}_{32} = -n_1 \hat{x} - n_2 \hat{y} + n_3 \hat{z} \tag{9.21}$$

$$\hat{r}_{321} = -n_1 \hat{x} + n_2 \hat{y} + n_3 \hat{z}$$

The resulting illumination area normal vectors and detected power levels for PD_3, PD_2, and PD_1 are

$$\vec{A}_3 = a^2 \hat{z}$$

$$\vec{A}_{32} = a^2 \left(1 - \frac{n_1}{2n_3}\right)\hat{y} \tag{9.22}$$

$$\vec{A}_{321} = a^2 \frac{n_2}{2n_3}\hat{x}$$

and

$$P_3 = -I_0\hat{r}_3 \cdot \bar{A}_3 = I_0 a^2 n_3$$

$$P_{32} = -RI_0\hat{r}_{32} \cdot \bar{A}_{32} = RI_0 a^2 \left(n_2 - \frac{n_1 n_2}{2n_3} \right)$$ (9.23)

$$P_{321} = -R^2 I_0 \hat{r}_{321} \cdot \bar{A}_{321} = R^2 I_0 a^2 \frac{n_1 n_2}{2n_3}$$

The six subcases described above are all defined for the directional cosine case of $n_1 < n_2 < n_3$, though each differs in its successive ordering of illumination and reflection. The ultimate photocurrents observed for each of the three PDs will be a result of all the derived possible reflection permutations that culminate in their respective PD area being illuminated. For example, the total incident power level detected on PD_1 will include five components:

1. The direct incident power onto PD_1
2. The reflected power due to a primary reflection off PD_2
3. The reflected power due to a primary reflection off PD_3
4. The reflected power due to a primary reflection off PD_2 followed by a secondary reflection off PD_3
5. The reflected power due to a primary reflection off PD_3 followed by a secondary reflection off PD_2

The resulting total power permutations are

$$P_{1-total} = P_1 + P_{21} + P_{31} + P_{231} + P_{321}$$

$$P_{2-total} = P_2 + P_{12} + P_{32} + P_{312} + P_{132}$$ (9.24)

$$P_{3-total} = P_3 + P_{13} + P_{23} + P_{123} + P_{213}$$

respectively, where the subscripts indicate the successive illumination and reflection sequence. The corresponding photocurrents, $i_1(t)$, $i_2(t)$, and $i_3(t)$, can then be found by summing the respective subcase power levels and applying a responsivity constant of proportionality, R, that describes both the internal PD quantum efficiency and optical transmissivity at the interface. For comparison, the PD photocurrents can be normalized with respect to $RI_0 a^2$ (representing photocurrents with normal optical incidence). For the first case of $n_1 < n_2 < n_3$, the photocurrents can be written as

$$i_1\left(n_1 < n_2 < n_3 \right) = n_1 + Rn_1 + R^2 \begin{cases} \left(2n_1 - \dfrac{n_1 n_2}{2n_3} - \dfrac{n_1 n_3}{2n_2} \right), n_3 < 2n_2 \\[2em] \dfrac{3n_1 n_2}{2n_3}, n_3 > 2n_2 \end{cases}$$

$$i_2(n_1 < n_2 < n_3) = n_2 + Rn_2 + R^2 \begin{cases} \left(2n_1 - \dfrac{n_1 n_2}{2n_3} - \dfrac{n_1 n_3}{2n_2}\right), n_3 < 2n_2 \\[3mm] \dfrac{3n_1 n_2}{2n_3}, n_3 > 2n_2 \end{cases}$$

(9.25)

$$i_3(n_1 < n_2 < n_3) = n_3 + R\left(n_1 + n_2 - \dfrac{n_1 n_2}{n_3}\right) + R^2 \dfrac{n_1 n_2}{n_3}$$

While the above PD$_1$, PD$_2$, and PD$_3$ photocurrents have been defined explicitly for the directional cosine case of $n_1 < n_2 < n_3$, the process employed in deriving (9.6) to (9.23) can be modified for the remaining five directional cosine cases. Rotational and mirror symmetries can be employed in this derivation, and the resulting three photocurrent expressions can be explicitly stated for each permutation. For the second incident light ray case, the x-, z-, and y-components of the incident light ray are in ascending order of magnitudes, and the inequality $n_1 < n_3 < n_2$ leads to three normalized photocurrents with

$$i_1(n_1 < n_3 < n_2) = n_1 + Rn_1 + R^2 \begin{cases} \left(2n_1 - \dfrac{n_1 n_3}{2n_2} - \dfrac{n_1 n_2}{2n_3}\right), n_2 < 2n_3 \\[3mm] \dfrac{3n_1 n_3}{2n_2}, n_2 > 2n_3 \end{cases}$$

$$i_2(n_1 < n_3 < n_2) = n_2 + R(n_1 + n_3 - \dfrac{n_1 n_3}{n_2}) + R^2 \dfrac{n_1 n_3}{n_2}$$

(9.26)

$$i_3(n_1 < n_3 < n_2) = n_3 + Rn_3 + R^2 \begin{cases} \left(2n_1 - \dfrac{n_1 n_3}{2n_2} - \dfrac{n_1 n_2}{2n_3}\right), n_2 < 2n_3 \\[3mm] \dfrac{3n_1 n_3}{2n_2}, n_2 > 2n_3 \end{cases}$$

For the third case, with $n_2 < n_1 < n_3$, the y-, x-, and z-components of the incident light ray are in an ascending order of magnitudes, and the three normalized photocurrents become

$$i_1(n_2 < n_1 < n_3) = n_1 + Rn_1 + R^2 \begin{cases} \left(2n_2 - \dfrac{n_1 n_2}{2n_3} - \dfrac{n_2 n_3}{2n_1}\right), n_3 < 2n_1 \\[3mm] \dfrac{3n_1 n_2}{2n_3}, n_3 > 2n_1 \end{cases}$$

$$i_2\left(n_2 < n_1 < n_3\right) = n_2 + Rn_2 + R^2 \begin{cases} \left(2n_2 - \dfrac{n_1 n_2}{2n_3} - \dfrac{n_2 n_3}{2n_1}\right), n_3 < 2n_1 \\ \\ \dfrac{3n_1 n_2}{2n_3}, n_3 > 2n_1 \end{cases}$$

(9.27)

$$i_3\left(n_2 < n_1 < n_3\right) = n_3 + R(n_1 + n_2 - \frac{n_1 n_2}{n_3}) + R^2 \frac{n_1 n_2}{n_3}$$

For the fourth case, with $n_2 < n_3 < n_1$, the y-, z-, and x-components of the incident light ray are in an ascending order of magnitudes. The resulting three normalized photocurrents for PD_1, PD_2, and PD_3 can then be written as

$$i_1\left(n_2 < n_3 < n_1\right) = n_1 + R\left(n_2 + n_3 - \frac{n_2 n_3}{n_1}\right) + R^2 \frac{n_2 n_3}{n_1}$$

$$i_2\left(n_2 < n_3 < n_1\right) = n_2 + Rn_2 + R^2 \begin{cases} \left(2n_2 - \dfrac{n_2 n_3}{2n_1} - \dfrac{n_1 n_2}{2n_3}\right), n_1 < 2n_3 \\ \\ \dfrac{3n_2 n_3}{2n_1}, n_1 > 2n_3 \end{cases}$$

(9.28)

$$i_3\left(n_2 < n_3 < n_1\right) = n_3 + Rn_3 + R^2 \begin{cases} \left(2n_2 - \dfrac{n_2 n_3}{2n_1} - \dfrac{n_1 n_2}{2n_3}\right), n_1 < 2n_3 \\ \\ \dfrac{3n_2 n_3}{2n_1}, n_1 > 2n_3 \end{cases}$$

For the fifth case, with $n_3 < n_1 < n_2$, the z-, x-, and y-components of the incident light ray are in an ascending order of magnitudes, and the normalized PD_1, PD_2, and PD_3 photocurrents become

$$i_1\left(n_3 < n_1 < n_2\right) = n_1 + Rn_1 + R^2 \begin{cases} \left(2n_3 - \dfrac{n_1 n_3}{2n_2} - \dfrac{n_2 n_3}{2n_1}\right), n_2 < 2n_1 \\ \\ \dfrac{3n_1 n_3}{2n_2}, n_2 > 2n_1 \end{cases}$$

$$i_2\left(n_3 < n_1 < n_2\right) = n_2 + R\left(n_1 + n_3 - \frac{n_1 n_3}{n_2}\right) + R^2 \frac{n_1 n_3}{n_2}$$

(9.29)

$$i_3\left(n_3 < n_1 < n_2\right) = n_3 + Rn_3 + R^2 \begin{cases} \left(2n_3 - \dfrac{n_1 n_3}{2n_2} - \dfrac{n_2 n_3}{2n_1}\right), n_2 < 2n_1 \\ \\ \dfrac{3n_1 n_3}{2n_2}, n_2 > 2n_1 \end{cases}$$

For the sixth and final case, with $n_3 < n_2 < n_1$, the z-, y-, and x-components of the incident light ray are in an ascending order of magnitudes. The resulting three normalized photocurrents are

$$i_1(n_3 < n_2 < n_1) = n_1 + R\left(n_2 + n_3 - \frac{n_2 n_3}{n_1}\right) + R^2 \frac{n_2 n_3}{n_1}$$

$$i_2(n_3 < n_2 < n_1) = n_2 + Rn_2 + R^2 \begin{cases} \left(2n_3 - \dfrac{n_2 n_3}{2n_1} - \dfrac{n_1 n_3}{2n_2}\right), n_1 < 2n_2 \\[2ex] \dfrac{3n_2 n_3}{2n_1}, n_1 > 2n_2 \end{cases} \qquad (9.30)$$

$$i_3(n_3 < n_2 < n_1) = n_3 + Rn_3 + R^2 \begin{cases} \left(2n_3 - \dfrac{n_2 n_3}{2n_1} - \dfrac{n_1 n_3}{2n_2}\right), n_1 < 2n_2 \\[2ex] \dfrac{3n_2 n_3}{2n_1}, n_1 > 2n_2 \end{cases}$$

The PD_1, PD_2, and PD_3 photocurrents above are summarized in Table 9.1 for the six permutations of the light ray directional cosine component inequalities. These analytic expressions have been confirmed with a brute-force MATLAB® ray-tracing program (not shown), in which the structure is illuminated by a fine mesh of light rays, and the three photocurrents are recorded for all possible incident angle combinations.

The three normalized theoretical photocurrents, i_1, i_2, and i_3, from (9.25) to (9.30), are displayed in Figure 9.3a to c. The photocurrents are plotted as a function of the azimuthal angle φ and polar angle θ. Notice from these figures that the photocurrent maxima for each of PD_1, PD_2, and PD_3 correspond to situations in which the respective PD surface is orthogonal to the incident light rays and the greatest amount of light is absorbed. Likewise, negligible photocurrents are observed when the respective PD surface is parallel to the incident light rays and the least amount of light is absorbed.

The directionality of the three PD photocurrents can be used for active control and optimization through a combination of differential sums. If, for example, PD_1 and PD_2 have the same photocurrent level, it can be expected that the incident light rays are balanced between the xz- and yz-planes. This alignment corresponds to the situation for which the optical source lies along the $\varphi = 45°$ plane. The same directionality arguments can be made by quantifying and balancing the differential signals between PD_2 and PD_3 and then PD_1 and PD_3.

The complete procedure for balancing and optimizing the photocurrent signals can be observed most easily by summing the three differential sum magnitudes. The result is shown as a surface in Figure 9.3d, where the photocurrent differential sum is defined as

$$i_{diff}(\varphi, \theta) = |i_3(\varphi, \theta) - i_2(\varphi, \theta)| + |i_3(\varphi, \theta) - i_1(\varphi, \theta)| + |i_2(\varphi, \theta) - i_1(\varphi, \theta)| \qquad (9.31)$$

TABLE 9.1

Theoretical Normalized Differential Photocurrents for PD$_1$, PD$_2$, and PD$_3$, Given Various Permutations of the Directional Cosine Components

Cases	PD$_1$ Normalized Photocurrents	PD$_2$ Normalized Photocurrents	PD$_3$ Normalized Photocurrents
$n_1 < n_2 < n_3$	$n_1 + Rn_1 + R^2 \begin{cases} \left(2n_1 - \dfrac{n_1n_2}{2n_3} - \dfrac{n_1n_3}{2n_2}\right), n_3 < 2n_2 \\[2ex] \dfrac{3n_1n_2}{2n_3}, n_3 > 2n_2 \end{cases}$	$n_2 + Rn_2 + R^2 \begin{cases} \left(2n_1 - \dfrac{n_1n_2}{2n_3} - \dfrac{n_1n_3}{2n_2}\right), n_3 < 2n_2 \\[2ex] \dfrac{3n_1n_2}{2n_3}, n_3 > 2n_2 \end{cases}$	$n_3 + R\left(n_1 + n_2 - \dfrac{n_1n_2}{n_3}\right) + R^2 \dfrac{n_1n_2}{n_3}$
$n_1 < n_3 < n_2$	$n_1 + Rn_1 + R^2 \begin{cases} \left(2n_1 - \dfrac{n_1n_3}{2n_2} - \dfrac{n_1n_2}{2n_3}\right), n_2 < 2n_3 \\[2ex] \dfrac{3n_1n_3}{2n_2}, n_2 > 2n_3 \end{cases}$	$n_2 + R\left(n_1 + n_3 - \dfrac{n_1n_3}{n_2}\right) + R^2 \dfrac{n_1n_3}{n_2}$	$n_3 + Rn_3 + R^2 \begin{cases} \left(2n_1 - \dfrac{n_1n_3}{2n_2} - \dfrac{n_1n_2}{2n_3}\right), n_2 < 2n_3 \\[2ex] \dfrac{3n_1n_3}{2n_2}, n_2 > 2n_3 \end{cases}$
$n_2 < n_1 < n_3$	$n_1 + Rn_1 + R^2 \begin{cases} \left(2n_2 - \dfrac{n_1n_2}{2n_3} - \dfrac{n_2n_3}{2n_1}\right), n_3 < 2n_1 \\[2ex] \dfrac{3n_1n_2}{2n_3}, n_3 > 2n_1 \end{cases}$	$n_2 + Rn_2 + R^2 \begin{cases} \left(2n_2 - \dfrac{n_1n_2}{2n_3} - \dfrac{n_2n_3}{2n_1}\right), n_3 < 2n_1 \\[2ex] \dfrac{3n_1n_2}{2n_3}, n_3 > 2n_1 \end{cases}$	$n_3 + R(n_1 + n_2 - \dfrac{n_1n_2}{n_3}) + R^2 \dfrac{n_1n_2}{n_3}$

(continued)

TABLE 9.1 (CONTINUED)

Theoretical Normalized Differential Photocurrents for PD$_1$, PD$_2$, and PD$_3$, Given Various Permutations of the Directional Cosine Components

Cases	PD$_1$ Normalized Photocurrents	PD$_2$ Normalized Photocurrents	PD$_3$ Normalized Photocurrents
$n_2 < n_3 < n_1$	$n_1 + R\left(n_2 + n_3 - \dfrac{n_2 n_3}{n_1}\right) + R^2 \dfrac{n_2 n_3}{n_1}$	$n_2 + R n_2 + R^2 \left\{ \begin{array}{l} \left(2n_2 - \dfrac{n_2 n_3}{2n_1} - \dfrac{n_1 n_2}{2n_3}\right), n_1 < 2n_3 \\[2mm] \dfrac{3 n_2 n_3}{2n_1}, n_1 > 2n_3 \end{array}\right.$	$n_3 + R n_3 + R^2 \left\{ \begin{array}{l} \left(2n_2 - \dfrac{n_2 n_3}{2n_1} - \dfrac{n_1 n_2}{2n_3}\right), n_1 < 2n_3 \\[2mm] \dfrac{3 n_2 n_3}{2n_1}, n_1 > 2n_3 \end{array}\right.$
$n_3 < n_1 < n_2$	$n_1 + R n_1 + R^2 \left\{ \begin{array}{l} \left(2n_3 - \dfrac{n_1 n_3}{2n_2} - \dfrac{n_2 n_3}{2n_1}\right), n_2 < 2n_1 \\[2mm] \dfrac{3 n_1 n_3}{2n_2}, n_2 > 2n_1 \end{array}\right.$	$n_2 + R\left(n_1 + n_3 - \dfrac{n_1 n_3}{n_2}\right) + R^2 \dfrac{n_1 n_3}{n_2}$	$n_3 + R n_3 + R^2 \left\{ \begin{array}{l} \left(2n_3 - \dfrac{n_1 n_3}{2n_2} - \dfrac{n_2 n_3}{2n_1}\right), n_2 < 2n_1 \\[2mm] \dfrac{3 n_1 n_3}{2n_2}, n_2 > 2n_1 \end{array}\right.$
$n_3 < n_2 < n_1$	$n_1 + R\left(n_2 + n_3 - \dfrac{n_2 n_3}{n_1}\right) + R^2 \dfrac{n_2 n_3}{n_1}$	$n_2 + R n_2 + R^2 \left\{ \begin{array}{l} \left(2n_3 - \dfrac{n_2 n_3}{2n_1} - \dfrac{n_1 n_3}{2n_2}\right), n_1 < 2n_2 \\[2mm] \dfrac{3 n_2 n_3}{2n_1}, n_1 > 2n_2 \end{array}\right.$	$n_3 + R n_3 + R^2 \left\{ \begin{array}{l} \left(2n_3 - \dfrac{n_2 n_3}{2n_1} - \dfrac{n_1 n_3}{2n_2}\right), n_1 < 2n_2 \\[2mm] \dfrac{3 n_2 n_3}{2n_1}, n_1 > 2n_2 \end{array}\right.$

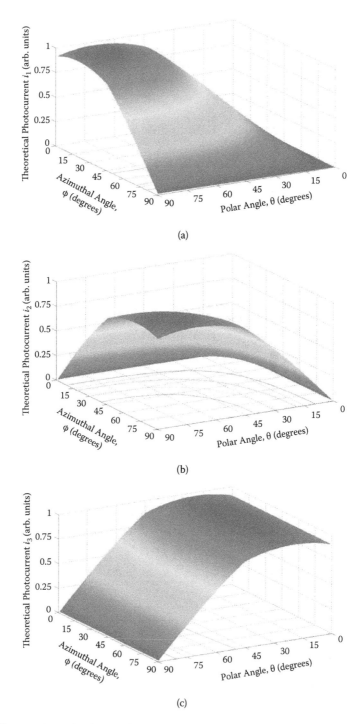

FIGURE 9.3 (See color insert.) The theoretical photocurrents (a) i_1, (b) i_2, and (c) i_3 are shown as surfaces varying with φ and θ. *(continued)*

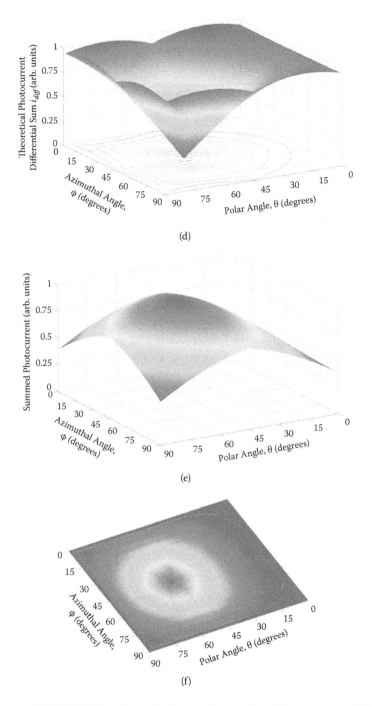

FIGURE 9.3 (CONTINUED) (See color insert.) The resulting (d) photocurrent differential sum and (e) absolute summed photocurrent are also shown. (f) Two-dimensional view of the retroreflected power.

Minimizing this photocurrent differential sum corresponds to an optimization of the angular alignment, with the balanced condition achieved when the absolute minimum of the displayed surface is reached. The systematic optimization will align the photosensor along $\varphi = 45°$ and $\theta = \cos^{-1}(1/\sqrt{3}) \approx 54.7°$. This balanced orientation is aligned for optimal communications between the optical source and photosensor.

From a signal detection standpoint, the summed photocurrent is recorded by combining the individual PD photocurrents. The absolute minimum of the photocurrent differential sum, shown in Figure 9.3d, leads to an absolute maximum of the summed photocurrent, shown in Figure 9.3e. This orientation gives optimal detection conditions. In contrast to this, differential photocurrent sum maxima at $(\varphi, \theta) = (0°, 0°)$, $(90°, 0°)$, $(0°, 90°)$, and $(90°, 90°)$ correspond to poor photodetection conditions for the summed photocurrent minima at these extreme orientations. The optimization procedure, leading to an orientation of $\varphi = 45°$ and

$$\vec{A}_{31} = a^2 \frac{n_3}{2n_2} \hat{x},$$

will bring the system into alignment for optimal signal detection.

From a signal retroreflection standpoint, the alignment optimization procedure will bring the system into an orientation for optimal retroreflection. The theoretical retroreflected power of this photosensor is shown as a surface varying with φ and θ in Figure 9.3f. Its signal strength is shown as a color map. It is readily apparent that the maximum retroreflected power (corresponding to the dark red regions) appears at the optimal orientation of $\varphi = 45°$ and

$$\vec{A}_{312} = a^2 \frac{n_1 n_3}{2n_2{}^2} \hat{y}.$$

In contrast to this, the dark blue regions around the periphery at $(\varphi, \theta) = (0°, 0°)$, $(90°, 0°)$, $(0°, 90°)$ and $(90°, 90°)$ correspond to near-zero retroreflection from the photosensor. Ultimately, the described differential triangulation and optimization procedure can bring the system into optimal alignment for bidirectional communication with enhanced photodetection and retroreflection.

9.4 APPLICATIONS

In the previous section, the theory for optical retroreflection and photodetection was developed for the proposed photosensor. The photosensor is now tested experimentally by way of OWC and OWL applications.

9.4.1 Optical Wireless Communication

The OWC system shown in Figure 9.1b is built and then tested with the photosensor. A broadband light-emitting diode (LED) source uniformly illuminates all the photosensor surfaces. A 5 m propagation distance is used. For angular testing purposes, the photosensor is mounted in a gyroscope that allows for independent rotations in the φ

and θ orientations. Phase-sensitive detection or electronic filtering can be employed as needed, to lock the PD signals into the modulation frequencies of interest. This filtering improves the sensitivity of the setup, removes extraneous/background optical signals, and compensates for the reduced reflectivity of the semiconductor surfaces (compared to, for example, metal mirror finishes). The cubic nature of the photosensor architecture also lends itself to arrayed detector distributions with interleaved photosensors distributed across surfaces. Such periodic arrays can be scaled up to increase the overall signal amplitude without sacrificing the individual PD response (RC) times.

To validate the differential photodetection process inherent to the corner-cube assembly, the photosensor is tested for angular characteristics. The PD_1, PD_2, and PD_3 photocurrents are recorded by a digital acquisition system as the azimuthal and polar angles are scanned in $10°$ increments over the ranges $0° \leq \varphi \leq 90°$ and $0° \leq \theta \leq 90°$. The results are shown in Figure 9.4a to c. The resemblance of these experimental results for PD_1, PD_2, and PD_3 to their theoretical counterparts in Figure 9.3a to c is immediately apparent. The results are in excellent agreement. The PD_1 photocurrent in Figure 9.4a rises from negligible levels at large azimuthal angles ($\varphi \approx 90°$) and small polar angles ($\theta \approx 0°$) to a maximum of $70\,\mu A$ when φ is small and θ is large. The PD_2 photocurrent in Figure 9.4b rises from negligible levels at small azimuthal angles ($\varphi \approx 0°$) and small polar angles ($\theta \approx 0°$) to a maximum of $68.6\,\mu A$ when φ and θ are both large. The PD_3 photocurrent in Figure 9.4c is largely independent of the azimuthal angle and rises from negligible levels at large polar angles ($\theta \approx 90°$) to a maximum of $72.9\,\mu A$ when θ is small. These rotational features are all apparent qualitatively by visualizing the PD illumination for angular extremes in the coordinate system of Figure 9.1a.

Having collected the independent photocurrents for PD_1, PD_2, and PD_3, the photocurrent differential sum can be calculated and used to optimize the alignment. The experimental photocurrent differential sum results are shown in Figure 9.4d as the azimuthal and polar angles are scanned in $10°$ increments over the ranges $0° \leq \varphi \leq 90°$ and $0° \leq \theta \leq 90°$. The resemblance of this surface to the theoretical curve defined by (9.31) and displayed in Figure 9.3d is clear. The theoretical photocurrent analyses and related assumptions, e.g., angular-independent surface reflectivity, appear to be validated, as the experimental data points in Figure 9.4d deviate from those of Figure 9.3d with a standard deviation below $10\,\mu A$. The observed deviations are greatest for orientations with large, i.e., glancing, angles of incidence off of one or more PDs. Such deviations, at the angular extremes, are well within the required accuracy of OWC systems.

The bidirectional OWC capability is verified here, given alignment along the optimized orientation with $\varphi \approx 45°$ and $\theta \approx 54.7°$. A $1\,mW$, a $650\,nm$ monochromatic laser diode is employed and expanded to create sufficiently uniform collimated beams directed at the photosensor. Both φ and θ angles are optimized with the aforementioned differential triangulation process. The $(x, y, z) = (1, 1, 1)$ orientation becomes aligned toward the laser source. Both active downlink and passive uplink schemes are demonstrated.

The active downlink mode is investigated first with the laser modulation at $2\,kHz$. The Pi-cell LC modulator is mounted at the entrance interface of the photosensor

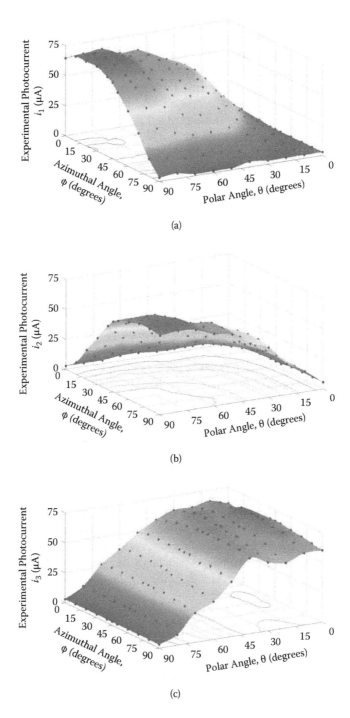

FIGURE 9.4 The experimental photocurrents (a) i_1, (b) i_2, and (c) i_3 are shown varying with φ and θ. *(continued)*

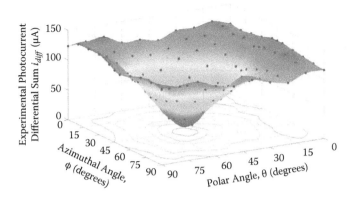

FIGURE 9.4 (CONTINUED) (d) Photocurrent differential sum.

device. The summed photocurrent, $i_1(t) + i_2(t) + i_3(t)$, is shown in Figure 9.5a as a function of time. The modulated characteristics of the laser source are seen to be effectively detected by the photosensor.

The passive uplink mode is also investigated. A biased silicon photodetector is located at the laser source to monitor the retroreflected signal. The photosensor is illuminated by the continuous-wave laser with the entrance Pi-cell modulator operating from a biased pi-state condition with a modulation frequency of 100 Hz. The resulting retroreflected signal level recorded by the photodetector is shown in Figure 9.5b as a function of time. Note that the Pi-cell LC modulator effectively maps the remotely encoded information onto the retroreflected beam. The signal level and modulation depth are seen to be appreciable with a signal-to-noise ratio (SNR) beyond 40 dB. This enhanced signal is due, in large part, to the previous beam alignment optimization procedure.

The photosensor of interest to this investigation is also tested with a broadband LED optical broadcasting scheme. A broadband LED local source is modulated at 4 kHz and broadcast toward the photosensor. After 4 m propagation, the light is reflected by the optimized photosensor and returned directly back to the local source, where it is sampled by a photodetector. The photocurrent signal recorded at the local source is shown in Figure 9.5c for the balanced orientation of $\varphi \approx 45°$ and $\theta \approx 55°$. The 4 kHz retroreflected signal is immediately apparent in the figure. For a broadcast optical power of 0.25 mW, a 10.6 nA local signal is detected with an SNR of approximately 40 dB. The low noise levels and large signal strength (well above the estimated 0.1 nA PD sensitivity) show the benefits of using an electronic bandpass filter stage, as it filters external noise sources and removes the contribution from background light levels. The same signal modulation can be seen at the remote photosensor in Figure 9.5d, where the PD_1, PD_2, and PD_3 photocurrent signals are displayed as a function of time. The remote signal levels here are well balanced, with an SNR of approximately 50 dB, leading to a corresponding local signal that is maximized and relatively insensitive to extraneous reflections in the environment.

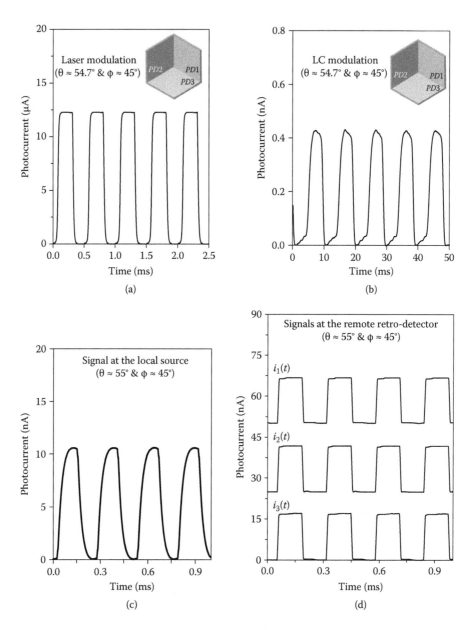

FIGURE 9.5 The photocurrents are shown for the (a) active downlink mode and (b) passive uplink mode, as a function of time, for the OWC system with laser illumination. The photocurrents are shown for the (c) local and (d) remote detection, with $i_1(t)$ (shifted up 50 nA), $i_2(t)$ (shifted up 25 nA), and $i_3(t)$. Both setups have the photosensor aligned at the optimal $\varphi \approx 45°$ and $\theta \approx 55°$ orientations.

9.4.2 OPTICAL WIRELESS LOCATION

In this subsection, the OWL capability of the proposed photosensor is presented experimentally. The photosensor is capable of estimating the angle of arrival (AOA), i.e., the azimuthal angle (φ) and the polar angle (θ), of transmitted signals from optical sources in an indoor environment by way of the differential photocurrents. The AOA information can then be used to estimate the position of the integrated photosensor by way of triangulation. In this case, the AOAs are used to define vectors, and the position is at the intersection of these vectors. At least two optical sources with known location are needed to do this. Such a system provides real-time indoor position information.

Proof-of-principle testing is then carried out with the indoor OWL system shown in Figure 9.1c. The schematic illustrates the triangulation process with two optical sources, i.e., LEDs and the proposed photosensor. Each LED specifies a line of position (LOP), i.e., vector of possible locations for the photosensor. Error in the AOA measurement causes the LOP to become a cone of possible positions. The area of possible photosensor positions is defined by the intersection of these two cones, as represented by the shaded area in Figure 9.1c. If a third optical source C exists, the cone emanating from that source would intersect the cones from optical sources A and B, and the intersecting area of all three cones would be reduced, resulting in a more accurate photosensor position. It is very important to note that the geometry of the optical sources, with respect to the position of the photosensor, affects the accuracy of the position estimate. This effect of geometry on positional accuracy, known as dilution of precision (DOP), can be quantified and is explained in Dempster.[24]

Positions in this indoor environment are referenced according to the (x_0, y_0, z_0) absolute reference frame, i.e., navigation frame. The photosensor is shown at the navigation frame origin ($x_0 = 0$, $y_0 = 0$, $z_0 = 0$) and aligned along the (x, y, z) relative reference frame, i.e., body frame. The photocurrents, $i_1(t)$, $i_2(t)$, and $i_3(t)$, are recorded for PD_1, PD_2, and PD_3, respectively, and comparisons between the incident signal levels are then made through the differential photocurrents $i_{13}(t) = i_1(t) - i_3(t)$ and $i_{23}(t) = i_2(t) - i_3(t)$, where it is noted that $i_{12}(t) = i_1(t) - i_2(t)$ provides no new differential information. These differential photocurrents, $i_{13}(t)$ and $i_{23}(t)$, are then used to estimate the AOA azimuthal angle φ and polar angle θ. For acquiring and processing differential photocurrents for multiple optical sources, a multifrequency LED emitter configuration is used.

The piecewise analytical expression for the photocurrent values, i_1, i_2, and i_3, as a function of angles φ and θ, was theoretically derived in Section 9.3 and experimentally verified in Section 9.4.1. In the OWL system, however, the operation is reversed. The PD photocurrents are measured and the AOA angles φ and θ are then estimated. To do this, the analytical expressions for photocurrents i_1, i_2, and i_3 are differenced to produce analytical expressions for the differential photocurrents i_{13} and i_{23}. These differential photocurrents are then inverted to produce expressions for φ and θ as a function of i_{13} and i_{23}. This is facilitated with polynomial approximations of the original function, containing a second-order term for φ and a third-order term for θ. A least angle regression algorithm[25] is then used. The root mean square (RMS) error

TABLE 9.2

Actual and Measured AOA Angle Errors for Use of the Photosensor with an Optical Source at Six Different Positions in the Indoor Environment

Position	φ Actual	φ Meas.	φ Error	θ Actual	θ Meas.	θ Error
1	6.9°	4.5°	2.4°	90°	89.3°	0.7°
2	21.3°	22.0°	0.7°	90°	90.0°	0.0°
3	45.3°	44.1°	1.2°	42.0°	41.9°	0.1°
4	55.2°	59.5°	4.3°	90°	89.1°	0.9°
5	69.3°	76.1°	6.8°	90°	90.4°	0.4°
6	71.9°	75.4°	3.5°	45.2°	42.3°	2.9°

of the approximation of both curves is less than 0.2%. In general, the i_{13} and i_{23} values can be used with this process to solve for the AOA angles φ and θ.

To determine the AOA estimation accuracy for various AOA values, the photosensor is fixed (in terms of position and orientation), and the photocurrents are collected to extract the AOA angles for an optical source at six different positions in an indoor environment. The actual and measured AOA φ and θ values, as well as the errors in φ and θ, are shown in Table 9.2. The RMS error in φ, across all six positions, is 3.8°, while the RMS error for θ is 1.3°. The RMS error for all angles combined (φ and θ) is 2.8°. The largest error occurs in the φ value for position 5. For all positions the error in φ is greater than the error in θ.

Assuming a system with K LEDs having fixed positions (x_i, y_i, z_i), $i = 1, 2, 3, \ldots, K$, results in K azimuthal angles φ and K polar angles θ for the photosensor at (x_0, y_0, z_0). The K azimuthal angles would result in K LOPs in the 2D (xy) plane between the photosensor and corresponding LED i. Solving for the gradient of any two LOPs would yield the estimated (x_0, y_0). Using the estimated (x_0, y_0), and measured θ to LED i, one can determine the elevation z_0 of the photosensor. Note that the largest RMS angular errors, i.e., $\sqrt{\varphi^2_{error} + \theta^2_{error}}$, occur for LED positions 5 and 6. With an LED at either of these two positions, the estimated photosensor position has an RMS error of 3.7 cm.

9.5 CONCLUSION

A new differential photosensor was introduced in this chapter through the merging of retroreflection and photodetection in an integrated optical package. It was shown that differential combinations of the photocurrent signals could be used as an active control mechanism to optimize the optical alignment. The theory for the optical retroreflection and photodetection characteristics was presented and subsequently verified experimentally. Excellent agreement was found between the theoretical and experimental angular characterizations. The photosensor was tested with OWC and OWL applications. The communication and location capabilities of the photosensor were found to be highly successful.

REFERENCES

1. Y. H. Lee and J. K. Lee, Passive remote sensing of three-layered anisotropic random media, in *Proceedings of IEEE IGARSS Conference*, Tokyo, 1993, vol. 1, pp. 249–251.
2. S. H. Yueh, R. Kwok, F. K. Li, S. V. Nghiem, W. J. Wilson, and J. A. Kong, Polarimetric passive remote sensing of wind-generated sea surfaces and ocean wind vectors, in *Proceedings of Oceans '93 Engineering in Harmony with Ocean*, Victoria, Canada, 1993, vol. 1, pp. 131–136.
3. P. Sharma, I. S. Hudiara, and M. L. Singh, Passive remote sensing of a buried object using a 29.9 GHz radiometer, in *Proceedings of Asia-Pacific Microwave Conference*, Suzhou, China, 2005, vol. 1, pp. 2–3.
4. R. D. Thom, T. L. Koch, J. D. Langan, and W. J. Parrish, A fully monolithic InSb infrared CCD array, *IEEE Trans. Electron Devices* 27, 1 (1980): 160–170.
5. C. R. Sharma, C. Furse, and R. R. Harrison, Low-power STDR CMOS sensor for locating faults in aging aircraft wiring, *IEEE Sensors J.* 7, 1 (2007): 43–50.
6. X. Y. Kong, GPS modeling in frequency domain, in *Proceedings of IEEE Second International Conference on WBUWC*, Sydney, Australia, 2007, pp. 61–61.
7. Y. Zhou, J. Schembri, L. Lamont, and J. Bird, Analysis of stand-alone GPS for relative location discovery in wireless sensor networks, in *Proceedings of IEEE CCECE Conference*, St. John's, Canada, 2009, pp. 437–441.
8. J. K. Lee and J. A. Kong, Active microwave remote sensing of an anisotropic random medium layer, *IEEE Trans. Geosci. Remote Sens.* GE-23 (1983): 910–923.
9. G. R. Allan, H. Riris, J. B. Abshire, X. L. Sun, E. Wilson, J. F. Burris, and M. A. Krainak, Laser sounder for active remote sensing measurements of CO_2 concentrations, in *Proceedings of IEEE Aerospace Conference*, Big Sky, MT, 2008, pp. 1–7.
10. M. A. Zuniga, T. M. Habashy, and J. A. Kong, Active remote sensing of layered random media, *IEEE Trans. Geosci. Remote Sens.* 17, 4 (1979): 296–302.
11. M. Steinhauer, H. O. Ruo, H. Irion, and W. Menzel, Millimeter-wave-radar sensor based on a transceiver array for automotive applications, *IEEE Trans. Microw. Theory Tech.* 56, 2 (2008): 261–269.
12. H. Dominik, Short range radar—Status of UWB sensors and their applications, in *Proceedings of IEEE European Microwave Conference*, Munich, Germany, 2007, pp. 1530–1533.
13. T. Dorney, J. Johnson, D. Mittleman, and R. Baraniuk, Imaging with THz pulses, in *Proceedings of IEEE International Conference on Image Processing*, Rochester, NY, 2002, vol. 1, pp. 764–767.
14. R. W. Mcgowan, R. A. Cheville, and D. Grischkowsky, Direct observation of the Gouy phase shift in THz impulse ranging, *Appl. Phys. Lett.* 76, 6 (2000): 670–672.
15. B. Warneke, M. Last, B. Liebowitz, and K. S. J. Pister, Smart dust: Communicating with a cubic-millimeter computer, *Computer* 34, 1 (2001): 44–51.
16. L. Zhou, J. M. Kahn, and K. S. J. Pister, Corner-cube retroreflectors based on structure-assisted assembly for free-space optical communication, *J. Microelectromech. Syst.* 12, 3 (2003): 233–242.
17. D. C. O' Brien, W. W. Yuan, J. J. Liu et al., Optical wireless communications for micromachines, *Proc. SPIE* 6304 (2006): 63041A.
18. W. S. Rabinovich, R. Mahon, H. R. Burris, et al., Free-space optical communications link at 1550 nm using multiple-quantum-well modulating retroreflectors in a marine environment, *Opt. Eng.* 44, 5 (2005): 056001–056003.
19. D. A. Humphreys and A. J. Moseley, GaInAs photodiodes as transfer standards for picoseconds measurements, *IET Proc. J. Optoelectron.* 135, 2 (1988): 146–152.
20. D. H. Auston, Picosecond optoelectronic switching and gating in silicon, *Appl. Phys. Lett.* 26 (1975): 101–103.

21. R. T. Howard, A. F. Heaton, R. M. Pinson, and C. K. Carrington, Orbital express advanced video guidance sensor, in *Proceedings of IEEE Aerospace Conference*, Big Sky, MT, 2008, pp. 1–10.

22. A. Makynen and J. Kostamovaara, Optimization of the displacement sensing precision of a reflected beam sensor in outdoor environment, in *Proceedings of 21th IEEE IMT Conference*, Como, Italy, 2004, vol. 2, pp. 1001–1004.

23. C. M. Collier, X. Jin, J. F. Holzman, and J. Cheng, Omni-directional characteristics of composite retroreflectors, *J. Opt. A. Pure Appl. Opt.* 11 (2009): 085404.

24. A. G. Dempster, Dilution of precision in angle of arrival positioning systems, *Electron. Lett.* 42 (2006): 291–292.

25. B. Efron, T. Hastie, I. Johnstone, and R. Tibshitani, Least angle regression, *Ann. Stat.* 32 (2004): 407–499.

Index

Milton Keynes UK
Ingram Content Group UK Ltd.
UKHW040108071024
449327UK00019B/907